Preventing and Handling Product Liability

Preventing and Handling Product Liability

Randall L. Goodden

Marcel Dekker, Inc. New York • Basel • Hong Kong

Library of Congress Cataloging-in-Publication Data

Goodden, Randall L.
Preventing and handling product liability / Randall L. Goodden.
p. cm.
Includes index
ISBN: 0-8247-9681-0 (alk. paper)
1. Products liability—United States. 2. Product safety—United States. 3. Risk management—United States. 4. Quality control. I. Title
KF1296.Z9G66 1995
346.7303'8—dc20
[346.73038] 95-33240
CIP

This publication was designed to provide helpful recommendations and assistance with regard to the subjects covered. It was not meant to supply direction on how to handle legal cases or issues, nor is it intended to take the place of legal advice or other expert assistance that a company may be required to seek.

The publisher offers discounts on this book when ordered in bulk quantities. For more information, write to Special Sales/Professional Marketing at the address below.

This book is printed on acid-free paper.

Marcel Dekker, Inc.
270 Madison Avenue, New York, New York 10016

Current printing (last digit):
10 9 8 7 6 5 4 3 2 1

PRINTED IN THE UNITED STATES OF AMERICA

To my wife, Peggy, for her love and support,
and my pride in life, Wendy and Becky.

Preface

The entire preoccupation with product liability in the last two decades, along with the time and money spent in litigation, are out of control in this country. The number of lawsuits related to product liability filed in federal courts increased 983% from 1974 to 1988, placing a heavy financial burden on corporations of all sizes. Many corporations have been required to pay substantial risk insurance premiums or defense costs, and what might have been healthy organizations have been forced into bankruptcy. Expenditures for product liability insurance premiums and other defense costs and settlements impact the bottom line and drain funds that would have otherwise gone into profits, and yet they are often viewed as a necessary evil unlike other Quality Costs. The impact of this has in many cases reduced the amount of funds available for research and development, preventing many new products from making it to the marketplace and adding significant cost to many other products.

As the number of attorneys in the United States grows, with over 40,000 entering the field every year and approximately 740,000 in total, and with record-setting awards to plaintiffs being publicized routinely, it is no wonder that such actions are almost instantaneous. America is the most litigious nation in the world, and, as of 1995, there is little relief in sight. Even with liability reform, it is doubtful whether such actions will ever be truly curtailed, because attorneys for plaintiffs will likely find new angles to pursue.

The insurance industry maintains that it spends as much on defending corporate policyholders against product liability actions as on paying claims. As this trend continues, insurance companies are focusing their efforts on controlling the costs of legal defense, and may often pursue an early settlement even if a case lacks credibility. Corporations and their insurance carriers must therefore join forces in a concerted effort to reduce risk and mitigate potential suits to turn this situation around.

The problem most manufacturing corporations have had up to this point is that they feel they really don't have the background or knowledge necessary to get involved in product liability. Many of those in Quality and Engineering management may have maintained concern and interest in the subject, but felt that a legal degree is necessary to get involved in this area. But this isn't true; the basic knowledge can be obtained without attending law school: one must only possess the desire to learn.

Many people acquire a substantial knowledge of medicine and the medical field without attending medical school, by reading and studying and/or through personal involvement. They do it because they want to take good care of themselves and their families, and they want to understand what they should and shouldn't be doing to maintain their good health. Although they're not experts, they can become well-versed.

The elements of product law can be learned the same way. Although the corporate representative designated to study this field will probably never become a lawyer, he or she will be able to improve and maintain the health of the organization. The first requirements are ambition, willingness, and recognition that one needs to learn, and then, with study and experience, the knowledge will grow.

This book demonstrates why corporations must abandon the conventional practice of leaving the handling of potential product liability incidents to the insurance companies, because it is obvious that the insurance companies (even with any experts they enlist) will never have the technical product knowledge that the corporation possesses. It is time for corporations to realize the inefficiency and risk in leaving their fate in the hands of others who lack the technical knowledge to challenge the credibility of the basic premise of a case, and get actively involved in their own defense.

This book will show any company how to develop the procedures and programs necessary to help reduce the potential of a product liability incident. It describes the Quality, reliability, and testing procedures and efforts that need to be put into place. It will show corporations how their own nonlegal staff can take a lead role in fighting—and winning—product liability actions. It describes the fundamentals and leading methods of investigating a field incident that might result in a product liability lawsuit, and stop it in its tracks. This book removes the mystique surrounding the legal process, and explains the procedures in terms anyone can understand.

The product liability field, and its associated costs and concerns, is a new frontier for management and the Quality professional. It is loaded with opportunities and rewards that many executives don't even know exist. It is a land mass that most corporate ships sail right past as though there weren't anything to be gained by landing here. It is a financial expenditure that is quietly accepted, and so buried in most financial reports that if it weren't for the attention that it periodically commands of CFOs it wouldn't get any attention at all. And yet the expenditure has an impact on profitability comparable to that of some of the leading elements of loss—such as scrap, rework, and warranty costs—that command all sorts of attention.

This book addresses a whole new area of focus for Quality professionals and engineering executives interested in getting involved and learning the improvement tactics that are capturing the attention of leading corporations everywhere, making them the pioneers in this new frontier.

Randall L. Goodden

Contents

Preventing and Handling Product Liability

1

Liability Statistics

THE TARGET: U.S. MANUFACTURERS

The fact that we are living in a very litigious age is more true now than at any other time in history, and more relevant in this country than in any other. And one of the largest targets for liability litigation is the manufacturing corporation. It is not uncommon for a manufacturing corporation with $75 million in annual sales to be involved in 10 to 30 or more cases of product liability, with larger corporations having enough to support a full-time legal staff.

What continues to prompt this action is the growing number of attorneys entering the field along with the publicized record rewards. The perception these attorneys seem to unanimously hold is that corporations and their insurance carriers have deep pockets to take advantage of, and that they would rather settle such situations early than fight them in court or face negative publicity. For the plaintiffs, it presents opportunity for gaining wealth with little effort or, from a lesser perspective or self-justification, it could be implied that it has arisen from society's own demand for safer products. But the fact of the matter is that the situation isn't likely to get any better without some radical changes in the law, or in the way we function as a society.

Increasing the odds against large manufacturers in trial is the current trend in corporate America to downsize or *reengineer* the organization. Many potential jurors could easily be victims of such corporate reorganization, or at least have feared such corporate moves, and in any case have grown to distrust corporations in general and develop a compassion for their coworkers.

The November 13, 1991, *Wall Street Journal* reported the results of a survey they had conducted, noting, "A large majority of jurors are predisposed to believe an individual's version of events in any dispute with a corporation." Corporations are no longer trusted as they once may have been when, generation after generation, workers developed loyalties to large companies in exchange for lifetime employment. Now corporations no longer exhibit commitment to employees, even though they may have been very dedicated individuals.

INSURANCE STATISTICS

In 1991 it was reported by the insurance industry that $1.6 billion was paid out in product liability losses. They also reported that an additional $1.1 billion was spent on legal costs in defending policy holders against claims—nearly 70% of incurred losses—including lawyer fees, court costs, costs associated with expert witnesses, and any other costs incurred for the defense (Figure 1). Ten years earlier the insurance industry reported that associated defense costs were 54% of the amount paid in losses.

In asbestos cases alone, which comprise the largest portion of product liability suits, 75% of the money paid out by manufacturers was to pay legal costs. The insurance industry also spent $1.1 billion on defense for medical malpractice cases, amounting to 40% of the $2.7 billion in losses. These two categories had the highest ratios of defense costs to losses of all types of liability casualty lines including auto liability, workers compensation, and other general liability. The bottom line is that the cost of defending commercial policyholders in product liability actions is rising faster than the losses themselves. All this is taken into consideration in the insurance premium paid by the corporation.

Ironically, what the corporate client would consider a case won the risk carrier may consider a major loss, because of the defense costs incurred. If a case was fought for 2 years, costing $50,000 in legal and expert witness fees, and in the end the plaintiff settled for $5000, the corporation could be disappointed that there even had to be a settlement but claim victory in that it was only $5000 when the plaintiff originally demanded $100,000. But to the risk handler it was a $55,000 loss.

In a study of legal fees related to product liability cases in 1991, loss adjustment expenses for cases totaled $1.6 billion, of which 85% was paid to defense attorneys. This would calculate out to about $908 million. Assuming that counsel for the plaintiff earned one-third of any award, the victim's attorneys would earn $521 million. In the end, this would mean that, for every dollar earned by counsel for the plaintiff, $1.74 was paid to counsel for the defense. Even though counsel for the defense isn't paid a percentage as counsel for the plaintiff is, it has to be remembered that the plaintiffs lose approximately 50% of their cases; therefore their attorneys wouldn't earn anything, whereas counsel for the defense is assured their hourly fees and expenses.

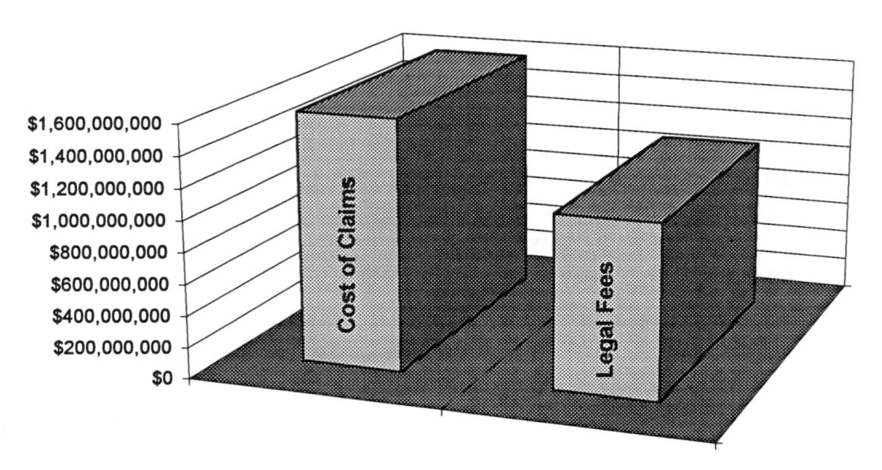

Figure 1 Impact of product liability cost on the insurance industry.

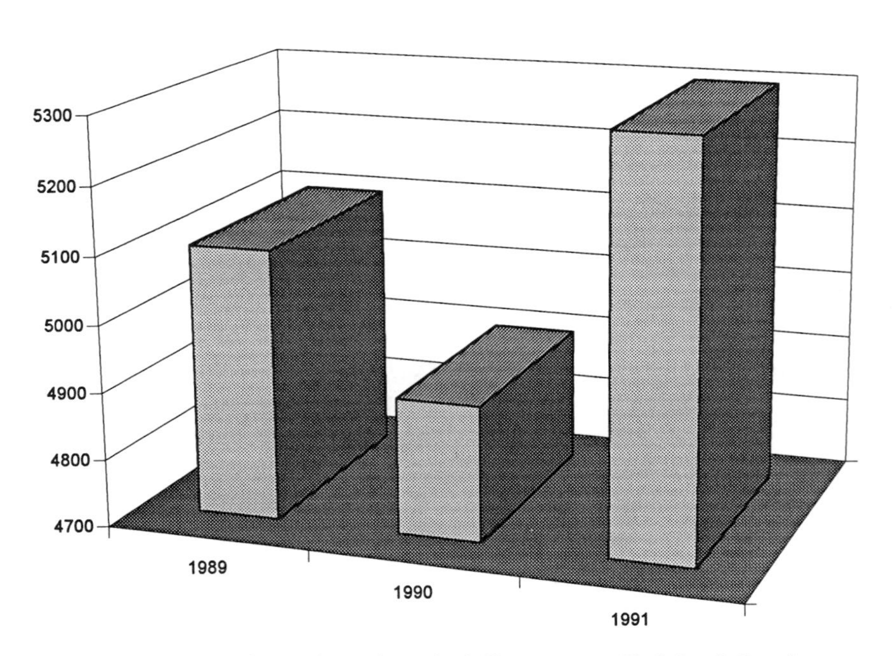

Figure 2 Number of product liability cases filed in federal court 1989–1991.

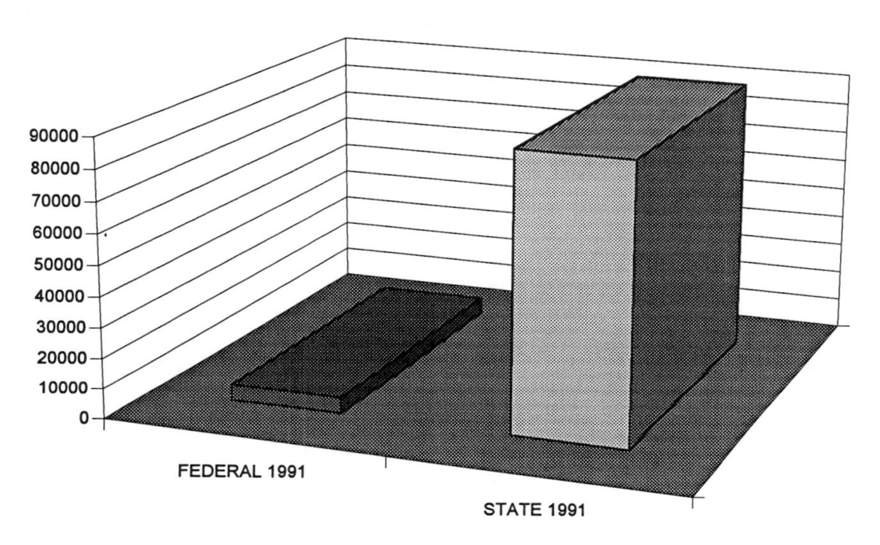

Figure 3 Cases filed in federal versus state courts.

Although some product liability incidents are settled even prior to entering into litigation, and a substantial amount of others are settled during the litigation stage with fewer than 4% actually going to trial, records still show that personal injury product liability cases (non-asbestos-related) that were handled in federal court rose from 5,100 in 1989 to 5,300 in 1991 (Figure 2). It is estimated that in addition to the 5,300 actions filed in federal courts, an estimated 90,000 were filed in state courts (the exact figure filed in state courts is unknown because only a few states keep separate numbers for civil cases) (Figure 3). If we add to the amount of cases filed in federal court those that were asbestos-related, our number of cases filed in federal court swells to 18,679 in 1990 alone. It should be noted that asbestos-related cases jumped 322% from 1985 to 1990, but non-asbestos-related cases fell 64% in federal courts during this same period. During the early 1990s, these non-asbestos product liability cases seemed to be on a rebound, especially in the state courts.

CHANCES OF WINNING

From a positive perspective, a study of product liability cases between 1989 and 1992 found that the chances of plaintiffs winning in a jury trial dealing with product liability in consumer products dropped from 59% to 39%. The improvement was less dramatic for industrial products, where the chances of winning declined from 65% to 57%. Another positive point to be made is that in all product liability cases that did end in some form of settlement up through 1992, only 1–2% of the cases were awarded more than $100,000. The average insurance payout per claim settled has fallen steadily from 1986, when, out of the total U.S. experience, the average payout was $18,190 (the highest point in a 10-year period), to 1991, when the average payout was $3,952 (Figure 4). An interesting side note: of those 1–2% of situations in which plaintiffs were awarded more than $100,000, the average jury award was $500,000, and in early 1993 the average award seemed to be rising.

In 1989 the General Accounting Office (GAO) studied 305 cases brought to trial in five states. The study found that:

The plaintiffs were awarded compensation in less than 50% of the cases.

The highest compensation awards were for total disability, which averaged $2.1 million; wrongful death, which averaged $937,000; and permanent partial disability, which averaged $524,000. They also found that the average award for temporarily disabling injuries was $78,000.

The average time between the initial complaint and trial was 2.5 years, and the average trial lasted 12 days.

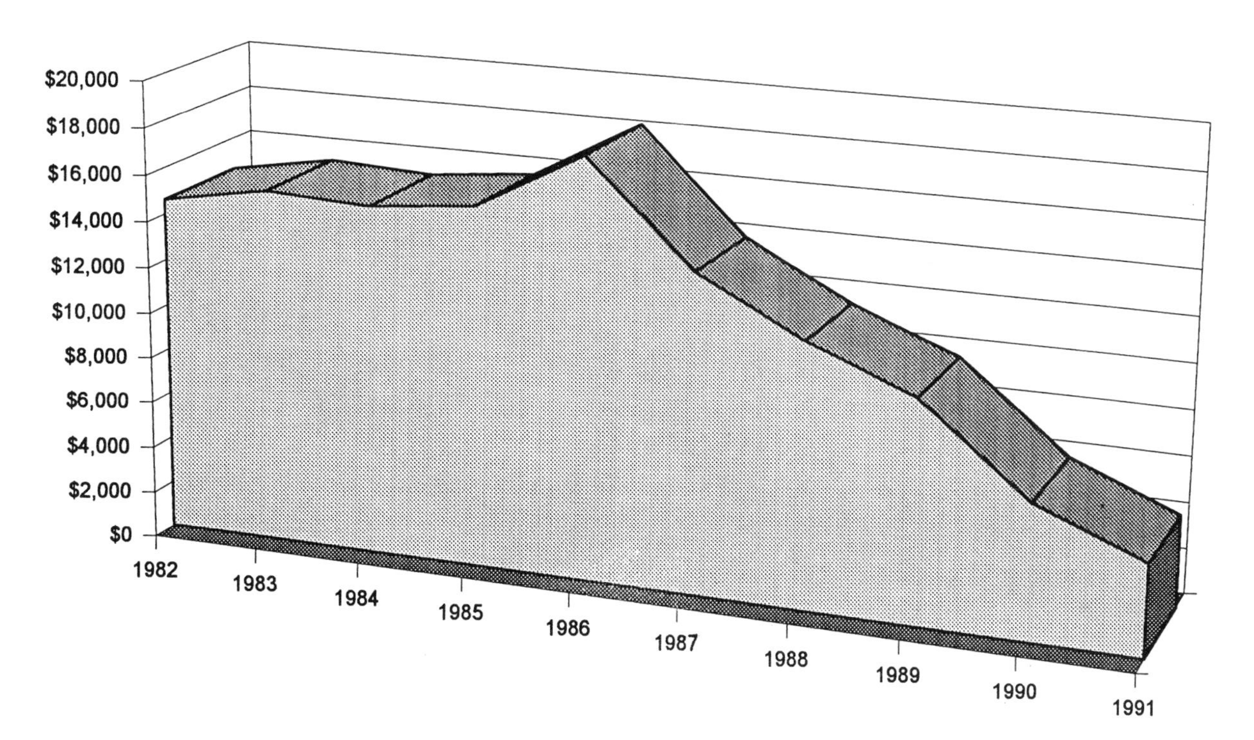

Figure 4 Average insurance payout in product liability cases—total U.S. experience.

The average plaintiff attorney received 33% commission, which amounted to $33,000; some earned over $1 million.
The average defense attorney was paid $20,000 per case, with the highest fee being $400,000.

It was also noted that upon appeal 71% of the original awards of over $1 million were reduced.

CAUSES OF PRODUCT LIABILITY

Another study, conducted in March 1993 by a major national insurance company, examined the losses involved in a random 27 product liability cases with settlements over $100,000. They found that the losses totaled $9,915,000, with the average being $360,000. They also found that the attributed causes were as follows (Figure 5):

Inadequate or nonexistent warnings (44%)
Inadequate guarding (26)
Design defects (21%)
Product defects (9%)

Furthermore, they found that 42% of the cases came from manufacturing companies, 28% from contractors, 19% from service industries, and 11% from retail establishments.

This activity has had a major impact on decisions by U.S. manufacturers to even bring products to the marketplace. A 1990 survey at an international conference composed of 3,600 organizations found that (Figure 6):

47% of U.S. manufacturers have withdrawn products from the market because of potential product liability concerns.
25% of U.S. manufacturers have discontinued various forms of product research and development.

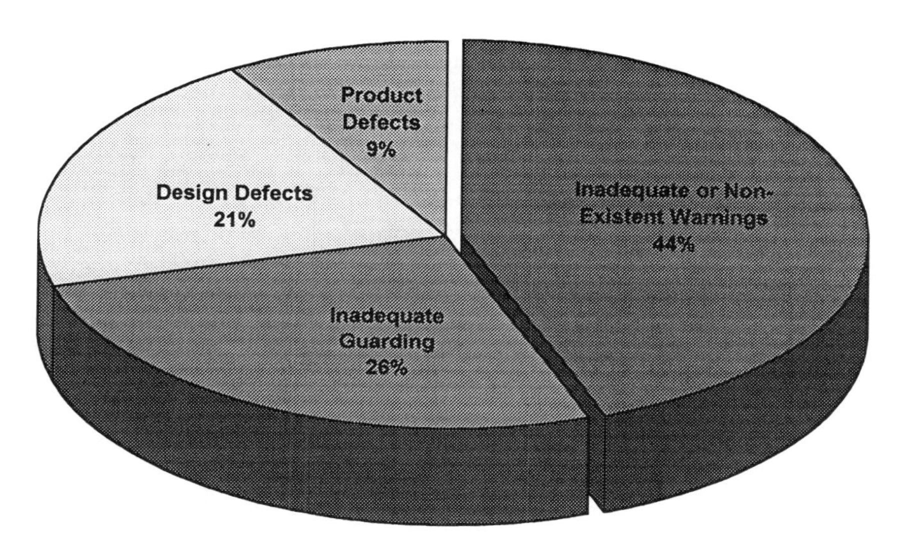

Figure 5 Breakdown of causes of product liability.

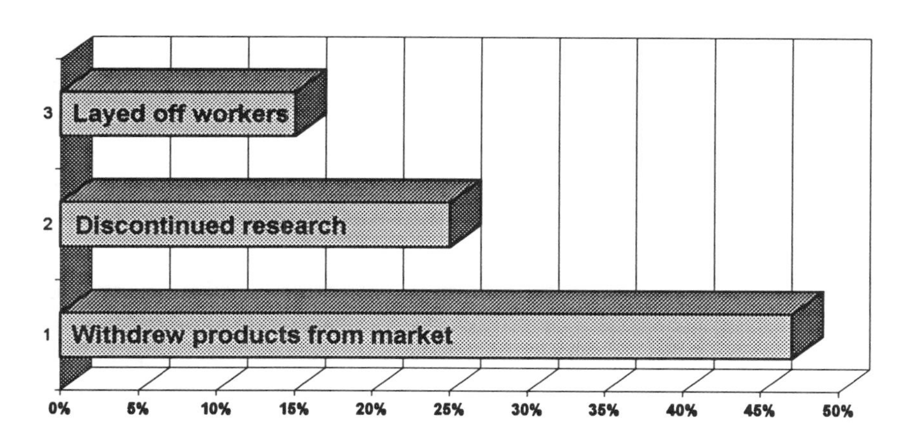

Figure 6 Impact of product liability on 3,600 manufacturers in 1990.

Approximately 15% of U.S. companies have laid off workers as a result of product liability actions.

Individuals, businesses, and governments spend more than $80 billion a year on direct litigation costs and higher insurance premiums.

In viewing such statistics, some might say that the fact that 25% of American manufacturers discontinued research into new products because of the potential of product liability is good, because the products may have been unsafe anyway. This is not a valid conclusion. The fact of the matter is that the legal system in the United States can so easily find fault with almost any product that some manufacturers just can't see the economic benefit in developing and marketing products that through consumer negligence and abuse may leave the manufacturer wide open for lawsuits. This growing exposure leaves American companies at a distinct disadvantage on a global scale as they compete in the markets of research and development but are effectively punished by the judicial system for developing new products and ideas.

It is also interesting to note, since we follow the Japanese lead in so many other areas of quality and product reliability, that Japan has no product liability laws to protect the consumer. Although they probably have an incident rate just as high as that in the United States, Japan was first beginning to look at introducing laws in this area in the fall of 1995.

The Japanese still have to protect themselves against product liability actions in the United States as well as in other countries. But even so, product liability costs have less impact on them than they do on American manufacturers. Although the number of cases has somewhat leveled, their costs may not have. The costs of product liability cases continue to increase with respect to litigation and defense fees. This is having a major effect on U.S. manufacturers, and on their ability to compete globally. It has been said that liability insurance costs are 15 times higher in the United States than they are in Japan, and 20 times higher than they are in most European countries. This has substantial impact on the prices and profitability of our products. Because of this situation, numerous national committees and organizations have been established to bring reform to the laws governing product liability. Two such groups are the Product Liability Coordinating Committee (composed of manufacturers, sellers, and various business leaders) and the American Tort Reform Association.

IMPACT ON PRODUCT COSTS

Although it is hard to identify the exact impact that product liability and the costs of related premiums have had on the products marketed throughout the country, a few examples have been publicized. For instance, one light plane manufacturer had to raise the cost of its plane $100,000 just to cover the costs associated with product liability coverage. In another example, 55% of the $100 pricetag of a typical football helmet is said to be the amount paid by its manufacturer in product liability premiums, which has kept many other potential helmet manufacturers from even entering the field. As a matter of fact, in 1970 there were about 20 football helmet manufacturers, and by 1990 there were just two that remained in business. In another product field, the typical cost of a certain standard vaccination soared from about $2.70 to almost $12 just to cover the costs of insurance.

Although these examples may depict product lines that one would think represent higher-risk areas, almost no product area is immune to such exposure. Take, for instance, our beloved Girl Scout cookies. The association reported in *Newsweek* (March 20, 1995) that it takes the profits of 87,000 boxes of cookies just to cover the costs of liability insurance for the Washington, D.C., area alone. This logically demonstrates that, in the end, we all pay more for what we buy because of the costs and risks involved in potential product liability.

THE PRODUCT LIABILITY DIRECTIVE IN THE EC

Just as the European Community (EC) unified their quality efforts and programs in the ISO Standard, they have unified their efforts in product liability in what is known as the Product Liability Directive. Basically, each EC member had its own laws on product liability, which complicated the concepts of free trade and unification. Back in 1976, the concept for a unified standard for product liability was proposed, but it was debated and resisted for many years. It wasn't until 1985 that the final Directive was issued, and even now a few countries may still not have totally bought into it and others have added a few of their own statutes and provisions. In theory, the Directive shifts the grounds for product liability from a fault-based perspective to strict product liability, which technically increases manufacturers' exposure.

However, in a number of other ways, the Directive along with the social atmosphere could still tend to discourage possible litigation. For instance, the fact that most of the member countries have some form of national health plan that would provide free medical care for an injured party would diminish the incentive or need for anyone to have to pursue such reimbursement. Second, and possibly more important, is the fact that the premise of a contingent fee is virtually unknown in most of the EC. So, the party who is contemplating litigation in compensation for such injury or property loss would need to ensure that the amount awarded would cover the legal expenses incurred. Furthermore, the losing party may also be responsible for paying his or her opponent's legal fees for having subjected them to such expense; if the opponent is a corporate defendant, such legal fees could be quite expensive. Or the plaintiff might win *part of* the case and lose other parts, or elements, of the initial complaint, and therefore could be responsible for some of the defendant's court costs. In one case, for instance, the plaintiff in a 100,000 Deutsche Mark claim was awarded DM 50,000 for the part of the claims about which the plaintiff was ruled to be in the right, but was forced to pay the defendant's legal costs for the other DM 50,000 part of the case that the plaintiff lost. To compound the difficulty of such a case, some of the courts in some countries either don't allow discovery or closely guard the discovery process.

This unified move should not present a barrier for U.S. manufacturers exporting products to the EC, for the product liability environment in the U.S. is by far more difficult. It should however, make U.S. manufacturers aware that they still need to ensure that the products they manufacture and ship overseas are safe and reliable, and not to view such a market as an opportunity to dump products that could not be sold domestically. Such a uniform code of laws should be looked at as being far more preferable than the climate in the United States, where each state has it own laws and guidelines on product liability, making it harder for the manufacturer to understand.

The U.S. lawmakers and courts should learn from what is happening in the EC with regard to this unified law, just as our manufacturers are learning from the introduction of the ISO standard in quality. We have far too many independent schools of thought in both fields, which makes it all the more difficult for us and our country to go forward. But, with as many firms and individuals getting rich off the system the way it is, there will always be significant opposition to change.

2

Risk Premiums and Quality Costs

The costs associated with risk coverage are normally handled by the CFO and the finance department, and known only to them and the CEO. Although the costs aren't always confidential, they probably aren't publicized within the company as much because few managers within the organization know anything about them, and there is a general attitude that there is little that anyone else can do about them anyway. Risk insurance becomes the necessary evil, and when the premiums get too high, the company begins shopping around for a different carrier.

Because of the lack of consideration regarding these premiums, the companies miss a major opportunity to save a great deal of money. They become one of the *Other Costs* in a financial review, or are lumped in with all the other business insurance costs, which are considered just part of the cost of doing business. They are expenses that Finance needs to focus on, as opposed to the rest of the executive staff, and no one—other than maybe the president—tends to question the figures.

It is interesting to note that the same corporation that pays little to no attention to this insurance premium—which could be costing $100,000 or more annually, depending on the company's size and history—will put every effort into combatting the causes of other costs such as those known as Quality Costs, or the Costs of Quality, which may be significantly lower in individual categories. A good corporation will dissect one of the Quality Costs, such as scrap, rework, or returns, and track on a monthly basis the amount that's being lost or spent, do a breakdown of where it is coming from, and statistically analyze the root causes in an effort to continuously improve in the area. Through dedicated internal efforts, many corporations are also very successful in bringing about continuous improvement in these areas as well as any other areas of added costs and losses, with the most improvements coming from the best efforts.

But, with respect to risk insurance and product liability, there is often no such tracking and control. Other than Finance, most departments have no idea what is even being spent, much less what the breakdowns are as they relate to any incident. In some cases, even the finance department may not be tracking actual costs and settlements being incurred by the insurance carrier. The carrier handles this area of concern, pays whatever costs it feels are necessary, and adjusts the corporation's premium accordingly upon renewal. So, when the corporation feels that the cost is rising too quickly, or if someone in Finance just wants to score points with the president, they pursue another insurance carrier that will handle their risk insurance at a lower rate—never minding the fact that the new insurer will probably just be offering this reduced premium as a lead-in.

The problem with this typical scenario is that these costs should be included in any other cost breakdown study that the corporation is trying to determine the cause of and address. A corporation dedicated to creating an atmosphere of continuous improvement, increased efficiency, reduced waste and cost, and thereby a higher degree of profitability, needs to be constantly tracking these expenditures along with those elements that normally comprise Quality Costs.

ELEMENTS OF QUALITY COSTS

Most leading companies already have a Quality system and effort in place to some degree. Those involved in the Quality effort should be using as its report card the various elements of Quality Costs, or the Costs of Quality or Poor Quality, to determine how successful their efforts have been and whether they are getting better or worse, and to quickly identify negative trends. Without this type of statistical tracking, companies can react only from gut feel, which normally has little accuracy.

Quality Costs are conventionally broken down into four basic categories: prevention, appraisal, internal failure, and external failure. These costs are then shown as a percent of sales, or percent of sales dollar volume of production. Technically, the four categories could have been broken down into two categories, showing only the amount of effort going into preventing problems versus the amount of failure coming out. If a company recognizes that it has a substantial dollar volume or percent of the whole in *failure* costs, and by comparison little money is being spent on *prevention*, they can intuitively recognize that more time and effort have to go up front to stop the amount of loss at the other end. The following is a definition of these categories.

Prevention

Prevention consists of the costs associated with trying to prevent failures. These are people, programs, and efforts initiated and dedicated to "doing it right the first time" and ensuring customer satisfaction. Examples are as follows:

Internal or external customer surveys.
Reviews of engineering drawings and specifications.
Testing and evaluation of reliability of new designs or components.
Supplier Quality Assurance programs and effort, including the time and expense of on-site evaluations and periodic performance reviews.
Qualification or validation of manufacturing processes and their control.
Quality planning, steering committees, and any other administrative time and expense.
The employee time and additional costs associated with Quality training and Quality awareness programs.
Statistical analysis work, SPC programs, and the tracking of other key performance indicators.
Quality performance reward programs or departmental improvement recognition programs.
The time associated with performing audits of the Quality system to ensure compliance.
Any other costs associated with trying to bring about Quality improvement to the organization.

Appraisal

Appraisal costs are those expended in the testing and evaluation of the products and processes initiated, such as audits and inspections, first piece approvals, reliability tests, and any other function that evaluates, or appraises, the quality of the work being performed. Examples of such costs are listed below:

Receiving inspection—the routine inspection of purchased materials or services.

The qualification of suppliers' material, such as the time and expense that would go into "first piece" analysis and approval.

Production audits and inspections, including process audits and final product inspections.

The expense incurred in job set-up costs or time spent testing or monitoring various processing pieces of equipment.

The expense of laboratory support or laboratory tests that are routinely performed on samples of products being produced.

The costs of various pieces of measurement equipment to ensure the quality of the product being produced.

Outside testing, analysis, or approval of the product being produced, possibly to attain certification or endorsement.

The costs associated with outside appraisal, inspections or audits of the product in the field.

Internal Failure

Internal failure costs are those incurred internally because a product or process failed to meet the quality requirements or specifications that had been established. Examples include:

Material that was scrapped out, including the accumulated increased value at the time it was scrapped.

Rework that transpired to bring a product up to an acceptable level as well as costs of bringing a supplier's product up to the level of acceptability, if the supplier wouldn't cover the costs.

The costs associated with troubleshooting and resolving problems found in the product or process.

The costs associated with engineering errors and the time taken to correct the discrepancies, along with any necessary rework or scrap incurred.

The costs of time needed to review nonconforming material and decide on the disposition.

External Failure

The category of *external failure* costs consists of all costs and losses encountered because the product or service provided to the customer was found to be unacceptable. This includes such items as:

All added costs associated with returned goods, including shipping and repair of product once returned.

The costs associated with warranty programs, including the costs of repairing the product in the field, repair or replacement parts, or even recalling the product if necessary.

The time and expense of investigating complaints received from the customer, but not including the actual costs contained under warranty.

And it is in this category that we would now place the costs of risk insurance premiums and begin tracking them.

This is a somewhat condensed list of what such costs may include, and expanded versions can be found in any book dealing with Quality Costs. The specific line items in each general category need to be identified by management; the costs associated need to be reported; and then management needs to track these actual costs on a regular basis and monitor the trends. But management needs to agree on the logic of the items they will track. Even though the above list is somewhat condensed, it may contain more items than a small company may what or need to track.

RECOGNIZING KEY QUALITY INDICATORS

As a Quality professional with about 20 years of experience, I have my own opinion as to what should be tracked in Quality Costs and what isn't necessary. For instance, I have found that the level of scrap, rework, warranty costs and returns presents both an adequate overall picture and a cross-sectional view of how well the company is doing. In a textbook version of Quality Cost breakdowns, each category would have 15 to 30 line items that would be tracked. This is fine for departments that have the time to devote to tracking so many areas, but those that don't should find that the abbreviated list will supply a decent sampling.

With the onset of TQM, far more items are loaded into the various categories, even though it becomes arguable as to whether such line items really have a relationship to true Quality—for example, the time spent in many administrative functions such as planning, troubleshooting or training, and in team meetings, functions, or activities focused on systems or operational improvement. This isn't to imply that these efforts shouldn't be taking place, because they should. The question becomes whether everything should fall under the Quality domain, or under the context of *corporate improvement*. It is also common for the prevention costs to include Quality salaries as one straight-line item.

The costs of failure should be recorded and monitored, and certain elements of *Prevention* and *Appraisal* can be informative as well. For instance, time spent by production personnel in inspection and testing could be looked at as added effort to ensure quality, and would be good elements to include because management should know how much time is being spent on these functions. The costs associated with training and manpower development as they relate to the Quality principles or fundamental tools are noteworthy items to log. It is not uncommon for large organizations to spend major dollars on training, and the costs to bring in outside consultants to do the training, without anyone really adding up the costs and at some point comparing them to the return. I have seen large companies spend as much as $500,000 in one year just on employee training programs, to achieve a higher level of quality.

But I don't feel Quality salaries—time spent by management in meetings and various types of strategic planning along with various other efforts that you know you should be doing—should be included by themselves. Quality salaries should not be lumped into *Prevention* but broken out into meaningful areas that required the time. For example, if a Quality employee spends time analyzing a customer field problem, then that time should go under *External Failure* as opposed to just under *Prevention*. You can see just by this example all the implications of properly classifying this time. But in some of the other areas, I think an organization can spend more time trying to track and log some specific elements of Quality Costs than what they're worth.

Furthermore, listing Quality salaries just as a prevention or appraisal cost implies that it is an added unnecessary cost, as though it should be decided whether it is a worthwhile investment. I feel Quality Costs should never be presented in this light. Although the Quality department should never grow to a disproportionate size, it should also never be reduced to a size that is no longer functional, or eliminated altogether. Failure costs are definitely unnecessary costs and should be avoided. When management makes an unusual investment in trying to reduce failure costs, that investment should be tracked under *Prevention*, and the results should be watched for under *Failure*. But it really doesn't take a very long list to determine whether a company is doing poorly or improving, and what areas require improvement, which is why the abbreviated list above may suffice.

REACTING TO PERCEPTIONS

The mistake would be for a company not to track any version of Quality Costs at all, and for management to react according to "gut" instincts that have no degree of accuracy at all. As an easy example, a company working under this style of management would react to every situation that created a lot of attention and would have little regard for the rest. For instance, if they suspect that they are losing a lot of money in scrap because they see scrap piled in every corner, they will want to determine what they should do to correct it.

Not knowing the root causes of the majority, or even what the majority is, they react to experience. Suppose that last week a material handler dropped a skid of product, which all came crashing to the floor because the handler had stacked the product in a careless manner. Because it attracted so much attention, management reacted and established a procedure on how, and how high, material can be stacked from this point on.

What they don't know is that a steady stream of scrap is being thrown out every day for some other defect. It doesn't draw any attention, because the daily amount does not indicate an alarming rate. But if they became aware of the figure on a monthly basis, they would learn that the loss is ten times that due to the material handling accident. So, unless a company records and statistically analyzes the various elements of Quality Costs, and breaks down the categories and line items into smaller increments in order to identify the major contributors, it is unlikely that they will have a good handle on controlling losses.

TRACKING PRODUCT LIABILITY COSTS

As noted, product liability costs would be recorded under *External Failure*. It could include, in addition to the dollars spent on premiums, added costs actually incurred due to a product liability incident. Use, for example, a situation in which a few hundred dollars was spent by a customer because of a small fire due to a corporation's defective product, and the corporation decided to pay the customer themselves rather than turn it over to their risk handler. Such an expenditure would go under *External Failure*. But it wouldn't include the possible $500,000 cost associated with a major fire or accident that is handled and paid for by the insurance carrier, even if it too was a result of poor quality.

As stated, the premiums paid for risk or liability insurance could go under this heading of *External Failure*, and the corporation would now track the trends of this item. If these premiums keep going up because of increased activity, costs and losses, attention should be drawn to it and some type of action should be taken. Likewise, the counterbalance would be the costs being invested by the company to prevent incidents of product liability from getting out of hand, such as the time and travel costs incurred during the initial investigation of an incident, once the corporation decides to follow the recommendations offered in this book. The objective, then, would be to determine if the time and money now being spent in *Prevention* is having any positive effect on the costs of premiums, or the amount being lost under *External Failure*.

For those companies that are self-insured for risk, every dollar spent in product liability settlements is immediately known and felt. In these situations, the executive staff that is tracking all the various financial cost reports should immediately see this expenditure and recognize the impact it has on the corporation.

For those companies that have millions in risk insurance coverage, the individual incidents do not normally impact the monthly premiums, but they will when it comes to contract renewal. The insurance carrier will keep a history of each case along a record of what was paid, and will base the contract renewal price on this history. Therefore, even when an incident doesn't immediately impact costs or expenditures, it will eventually have a long-term impact on costs.

UNDERWRITING PREMIUMS

A corporation's risk insurance premium with its current insurance company, or the premium likely to be quoted by a new company, is generally based on four elements. First, the insurance company, or the Loss Control Staff, will base the manufacturing corporation's premium on what they consider the manufacturer's exposure to risk, that is, the amount of sales the company is generating, and the number of products or units that the company is manufacturing or shipping. Naturally, from this perspective, the more individual products that are placed into commercial trade, the greater the potential risk. The Loss Control Staff would be looking for the answers to these questions:

What does the company manufacture and how long have they been in business?
What services does the company provide, or do they subcontract work?
Are their engineers and designers qualified for what they are doing?
What Quality controls do they have in place that govern suppliers through shipment?
What Quality records are being kept to substantiate the efforts undertaken?
What warranties or guarantees do they offer with their products?
What is the insurance carrier's loss potential?
Who are the manufacturer's customers, or end users of their products?
How does the manufacturer handle and control its marketing system involving literature and brochures? Is this material being reviewed from a legal perspective?

From there, the insurance underwriters consider the manufacturer's loss experience. The manufacturer must furnish the insurance company with this record, even if it is a new carrier. The loss experience, based on the sales volume, allows them to calculate a "loss rate," or expected number of losses per sales dollar or manufacturing volume.

The next element that goes into the calculation of the premium cost is the schedule debits or credits. Once the underwriters have established the base calculations, they then apply credits or debits based on whether they feel the manufacturer is above or below the national average in its efforts and abilities to prevent the possibilities of exposure. At this point the manufacturer will want to share with the insurance carrier information about all their new programs and efforts to prevent and eliminate potential product liability (as described in this book) once they are employed, in order to bring about a reduction in the premium.

Lastly, the insurance underwriters consider the insurance market conditions. This is where they track the insurance cycle to determine whether that period of time is considered a buyer's or a seller's market. This takes into account the current interest rate levels; if the interest rates are low, then the insurance company's investment income is normally less than when interest rates are high. Therefore, if they aren't receiving as much income from their own investments, the premiums tend to be higher. Also taken into account as part of this market study is the current availability of insurance. If there isn't much insurance being made available, it allows them to drive up the cost, whereas if a lot of insurance is available, the insurance companies have to compete against one another.

Independently, manufacturers can't have much influence over interest rates. And if they have a good product at a good price, their sales and volumes are only likely to continuously grow. So the only areas where the manufacturer can make a difference are in their *loss experience*, especially as the loss rate may compare to that of other manufacturers, and in the debits and credits for preventive efforts and programs that are being put into place. Once these efforts have begun to take effect, the manufacturer will begin to be able to show the insurance carrier and companies the lower rate of incidents that the company is experiencing over the recent past. Along with the benefits of all the new preventive efforts being made, the manufacturer should benefit substantially in the lowest possible premium cost.

Manufacturers shipping products into the European Community might consider obtaining their product liability insurance from overseas risk carriers, where the costs are normally lower than in the United States because of the more amenable product liability (as explained in Chapter 1), but also because a European carrier will be a much greater resource to an American manufacturer should an incident of product liability arise.

The roles and efforts introduced in this book are so new to industry, insurance and the legal field that the insurance carrier isn't even likely to inquire about such efforts in its initial or annual interview. They will probably inquire about Quality programs and efforts, because those are accepted and expected programs. But this type of preventive effort is in its infancy. Once the manufacturer implements the programs described in this book and takes the time to give a full presentation of the new programs to the insurance carrier, the carrier will be extremely impressed, which should lead to reduced premiums.

3

Conventional Risk Management

The average corporation handles risk insurance much like health insurance. They find the company that will give them the best price for what they want, and give them their business. If the premiums begin to cost too much after a few years, they will probably switch and go with another insurance company. Unlike the way in which the corporation is likely to handle other relationships, there isn't any customer–supplier relationship built here, nor any type of alliance developed. The corporation needs adequate coverage, and they're looking for the best possible price.

Finding insurance at a lower cost isn't an easy task. The incidents of liability become a matter of record that the corporation carries with it wherever it goes. Even though they may find a company that will insure them for less, it will still cost more than had they been able to maintain a clean slate. It is similar to a person's driving record: if a driver has an accident or is convicted of an offense, that record goes with him no matter where he goes for insurance, although he may find a company that will insure him for less than he was paying before.

As important and logical as it would seem that both parties (the corporation and the risk carrier) would fight for and protect the common good, it doesn't happen, primarily because there are two points of view: the good of the corporation and the good of the insurance carrier. Both have at least one interest in common: preventing dollar losses. But the corporation's other major interest is its reputation.

When an incident arises that falls into the insurance company's area of responsibility, the corporation expects them to handle it. The typical corporate attitude would be: "Let us know if there's anything we need to do or provide, but otherwise we trust that you'll handle this for us. After all, you're the experts in this area." The trouble is they're *not* the experts, and corporations need to start realizing that. Moreover, the insurance company's primary goal will be to minimize losses, not necessarily to try to save anyone's reputation.

When it comes to a product, its manufacturer knows more about it than anyone else, and should know more about its capabilities than the insurance company would ever hope to know. What the insurance company *does* know the most about are typical processes and procedures to be followed surrounding an accident or injury, and they can coordinate all the events. But they will never be rated as even rank amateurs when it comes to knowing about your product, and this is one of the reasons that corporations have to get more involved in product liability incident investigations. But because so many corporations don't get involved, it becomes somewhat of a challenge to do so.

A TYPICAL CASE

Here is a the normal sequence of events followed in a hypothetical product liability incident. The company in question manufacturers the standard promotional sign that is commonly seen in store windows across the country. They have been making hundreds of them for every major company for many years.

One day, they receive a notice from an attorney that a client he is representing was severely and permanently injured because one of their signs became defective. He states in a letter of complaint that they are demanding $300,000 in damages if they can settle this within 30 days. Beyond that period, the price will go up considerably as they enter into litigation and prepare for a court battle. The company is somewhat suprised by the notice, but wastes no time in contacting their insurance company and forwarding the paperwork to them.

The insurance company begins to follow its normal routine at this point, and requests from the attorney all the accident reports, hospital and medical files and physical evaluations, personal background of the injured party, witness accounts, investigative and expert reports, tax returns for the past 3 years, and anything else that exists surrounding the alleged incident or related to the alleged loss. Then they ask for an extension on the $300,000 offer. This probably takes about a month for the attorney's office to put together and send out, but they will do their best to ensure that all the records submitted will appear to substantially back the claim. Meanwhile, the original manufacturer involved doesn't hear anything with regard to what is going on and, taking a "no news is good news" attitude, trusts that the insurance carrier has the situation well under control.

Once the requested information is received, the insurance company will sift through it for another couple of months, and in the end send the attorney a letter stating that they would like to receive countless other reports and information regarding the injured party; they would like to discuss the evidence with the attorney and have the opportunity to ask some questions regarding the client's medical claims; and they might like to have an expert of their own analyze the product—the neon sign—and determine the credibility of the claim.

If the attorney is trying to put little cost and effort into this case, he will pursue it from the perspective of strict liability, or the product's being unreasonably dangerous for the marketplace or from lack of adequate warning. Otherwise, and if he hasn't done it already, the attorney will pursue an expert opinion to back up the claim that the product was indeed defective. This will be from someone who is local and doesn't have to spend much time determining that the product was defective in some manner, but the individual will appear by all standards to be an expert in the related field: someone such as an electrical engineer who works as a consultant but does quick little jobs like this for a few hundred dollars each, or a college professor moonlighting in this area. In any case, when the report is submitted, it will come across as an unchallengeable expert opinion in the case, loaded with technical and electrical elements that no other amateur will even understand, especially an insurance investigator.

The insurance company will also try to find someone locally to serve as their expert and review the attorney's report and determine the credibility, and possibly to look at the product itself. But the insurance company's biggest focus will be on the medical claims and claims of permanent injury. This is where the insurance company is able to call on a lot of experience and resources. By poring through the medical claims, and possibly asking for second opinions from their own doctors, the insurance company can drag the case out for many months. They may also elect to get into personal surveillance to try to disprove the plaintiff's claim of permanent injury.

In any case, the insurance company's focus will be less on the product, other than to verify that it indeed was yours, and more on challenging all the medical claims and disproving the extent of the injury or loss. All this can be accomplished by the insurance carrier's own people, which means that it really isn't costing them anything at this point, and they can drag this case out for months in the hopes of wearing out the attorney. The delay may not even be intentional: insurance companies work on numerous problems in all areas, so it is very common for specific situations to take months to rectify merely because of the caseload and response timing.

If the plaintiff's attorney, on the other hand, wants to force the insurance company into a quick settlement, the attorney will initiate a court action, thereby demonstrating that he fully intends to get the case into court and in front of a jury as quickly as possible. Early in the case, counsel for the plaintiff figures that he has a 50/50 chance of winning this case if he can get it in front of a jury, without his having to put much work into it. He also knows that the last thing the insurance company wants is to pay the costs of a court battle, so to some extent he is calling their bluff. He may therefore push to take his chances early rather than allow the case to carry on for months and years, inevitably forcing him to increase his investment in it.

If counsel for the injured party does pursue court action, the insurance carrier is forced into retaining counsel themselves, which they realize will be a significant added expense. So they retain the assistance of a local attorney, but they limit his activity as much as possible. Once legal counsel has entered the picture on behalf of the defendant (that is, the manufacturer), the case will really have a tendency to drag out. Between the time it takes for the insurance carrier to bring their new counsel up to speed and the time it takes for this defense counsel to start communicating with counsel for the plaintiff, more months will go by. One of the most aggravating things for the manufacturer, if they do take an active interest in the case, is that things take forever to happen.

One of the first reports that the insurance company gets from their newly hired attorney is an estimate of what his services will cost should the case be settled early, settled after some litigation, or run the full course. Additionally, they want to know as soon as possible what the attorney estimates as their chance of winning this case should it run the full course. The insurance company's only concern at this point is cost: the least expensive way to go and their chances of winning.

THE PRIMARY CONCERN: COST

Costs are almost always a key factor in this whole scenario. The way insurance companies make their money is a unique concept. Any other type of a company manufactures a product or provides a service, sells it to a customer, subtracts the costs incurred from what they were able to sell it for, and in the end hopes they made a profit.

With insurance companies, however, you basically pay them for nothing, and the money they receive from you is entirely profit. Then, when incidents occur, they begin to lose some of this money as they incur costs for handling those situations. And, since your premium was a fixed cost (at least for now), it means that they also have a limited amount of your money to work with. If they exhaust those funds, they then have to use their own money to continue covering costs. The whole situation now becomes a real loss to them.

In our hypothetical situation, where the opposing counsel has initiated legal action, all the things the insurance company does with their own attorney from this point on will be added cost, which obviously erodes potential profit. If they need to retain an expert to analyze the product, it will cost the insurance company. If they feel that the plaintiff needs to be questioned or deposed, it will cost the insurance company money. And naturally, if the case actually goes to court, it will cost a considerable amount. So they really try to control all these actions and costs as much as they can.

Of course, the attorney has already sent in an estimate of costs, so he too has to control his involvement.

Because of the insurance company's concerns about costs, the attorney must get their permission to do anything regarding the case prior to actually doing the work. This is obviously not in the corporation's best interests. But at this point the insurance carrier isn't really concerned about what would be best for their client, only what would be best for them from the perspective of what will cost the least. This is discussed further in Chapter 1, "Liability Statistics," in which I told how the costs for defending product liability are as much as the losses from the same. Knowing the problem that they are dealing with here, as shown by such statistics, will make insurance companies even more prone to curtail costs so the defense attorney must constantly justify every move, and prove to the insurance carrier why it is necessary and how they'll benefit from it.

THE MANUFACTURER'S BEST INTEREST

It is in the manufacturer's best interest to ensure that they don't lose such cases and that the cases are not settled out of court. The manufacturer's attitude should be that they will fight the case to the end in order to prove their innocence (unless, of course, they know that they aren't innocent . . . in which case they will still fight it). The manufacturer doesn't want such an incident on their record, because it can be very damaging in the future. Not only will it affect future premiums because it becomes a matter of record regardless of what insurance company you do business with, but, more importantly, such convictions are discoverable by attorneys representing future plaintiffs and in future cases, and such incidents will add a lot of fuel to the fire that the manufacturer has a history of such problems. If the manufacturer doesn't even get involved, however, they basically have no attitude toward this whole thing and the tone is set by whatever the insurance carrier's attitude is, or what they are the most interested in accomplishing.

In our hypothetical situation, the defense attorney is constantly forced to get permission from the insurance carrier to pursue any course of action that he sees as beneficial. All the while, the insurance carrier is adding up what this case is already costing them, and continues to ask the defense attorney to estimate future costs, and weighs the benefits of fighting the case any further versus pursuing some sort of a settlement. The corporation whose reputation and future are at risk remains oblivious to what is going on.

PURSUING AN EXPERT

As touched on before, one of the things that both parties inevitably have to do is seek out experts on the case and get their reports together on what may have gone wrong in this instance. To control costs, knowing that the person may have to be brought in a number of times for meetings, discussions, and depositions, this expert will be someone local, someone who would seem credible and should know what he is talking about but can be had cheaply. Someone either who has a long list of credentials if he is an independent consultant or who represents academia, and would therefore not be challenged. In short, someone who can give a pretty impressive report for about $300–500. But the manufacturer's own insurance company as well as their defense counsel wouldn't think about coming to the manufacturer in search of this help and expertise. This is what they are being paid to handle for the manufacturer, so they don't want to appear incapable of handling the job.

Another irony in this typical sequence of events is that, if the case does go to court and is heard by a jury of individuals who have little or no technical expertise, the outcome will depend laregly on who can provide the most convincing experts to testify on their behalf. And the irony is that the people most knowledgeable about the manufacturer's product and what it is capable of doing, and the ones who can speak the most intelligently about it, are at the corporation. Yet they may never be recruited to serve in this final battle, which is fought with rented soldiers whose weapons are merely their own theories or educated guesses.

THE FINAL OUTCOME

In the end, the insurance carrier merely notifies the manufacturer of the outcome and it becomes a matter of record. This is especially true if it ends in a settlement, which happens in 96% of cases, as opposed to if it went to trial, which would logically have to involve the defendant to some degree and certain employees. In an out-of-court settlement, if the other party was suing for $300,000 and the insurance company could instead get them to settle for $150,000, both parties appear to be victorious. But the fact is that the incident resulted in a $150,000 insurance expense, and that will be reflected in the premiums for the years that follow.

If the case does go to trial, and the corporation loses (possibly because of the expert witness who testified on their behalf), the corporation loses not only with respect to the award granted the plaintiff but also from the perspective that it has now added this loss to its discoverable record. This will make the next case even harder to win, and the next, and the next. And all the while the premiums will continue to rise and rise.

OTHER TYPES OF CASES

In the hypothetical situation described above, we dealt with an injury and how it would logically be investigated and handled. The fact that you are placing a value on something that isn't a material asset allows for that value to be challenged. But take a different scenario where, say, the product liability incident resulted in property loss. Let's say that instead of having caused an injury, the promotional window sign allegedly shorted out and caused substantial fire damage to the store. The fire chief of a small-town fire department comes in to determine cause, and writes down that the fire appeared to have started in the area of some electrical signs, possibly from a short or an electrical overload. The report is pretty vague, but very typical.

The store owner calls his insurance carrier and makes them aware of what happened, and is naturally interested in getting the place rebuilt as quickly as possible so he can get back into business. The insurance company sends over a claims representative to look at the damage and work up some sort of estimate. The investigator is also very interested in determining the cause, so he first asks the store owner for the fire report. With that in hand, he inquires about the signs that were referenced and looks at the point of apparent origin. The store owner shows the investigator the remains of the sign that was in the area where it is believed that the fire originated. Upon examination of the electric sign, the investigator sees the name of the major product company being advertised and hangs onto the sign.

Because the fire report is somewhat vague, the insurance investigator calls on the services of a small local forensics lab to (as in our previous scenario) obtain an expert opinion as to the cause. To save time (and expense), the insurance investigator brings the forensics inspector right up to where the sign was and states that this is pretty much what the fire department determined was the cause. The forensics inspector takes a quick look around and agrees that the fire seems by all indications to have started in the vicinity of the promotional sign, probably due to an electrical short in the product or some type of malfunction, and forwards a report to the insurance company with plenty of pictures and possibly a video. The product itself has been pretty much destroyed at this point, so nothing can be proven.

The store owner's insurance carrier immediately pays to have the place rebuilt and get the guy back in business. During this same period, the carrier pursues subrogation from the company advertised on the sign, which argues that the complaint must be made to the actual manufacturer and provides the name of the original corporation. The insurance company now forwards the letter to the corporation, where it is forwarded to the finance department, which once again forwards it to their liability insurance carrier.

Once again, the insurance company either accepts the fact that the product was at fault or tries to find their own local, low-cost expert to determine the possibilities. Although both parties could hire attorneys and take the matter to court, instead of having one attorney and one insurance company involved as in the personal injury case, we now have two insurance companies, and both parties recognize the added expense of legal battle. So they tend to shy away from it and resolve their differences, and hopefully come to some kind of agreement.

In addition, if the store owner is also trying to recoup lost earnings, which may or may not have been covered by the insurance, then the corporation's insurance carrier once again has something that they can really get their hands on, and investigate that aspect of the claim for months. But, as it relates to the original claim, they either go along with the allegation or decide to fight it.

These are the typical ways in which many such product liability issues are currently handled by insurance carriers, demonstrating why manufacturers need to become actively involved. Not only are the manufacturers the best experts on their own products, but they can save the insurance company a substantial amount of time and expense in evaluating any situation and determining the possibilities. Participating in the defense is also an excellent way for the manufacturer to gain more knowledge of how their products are used, as well as a lot of expertise in forensics work.

4

The Quality System

Quality is probably the single most focused area of effort in corporate America today. More books, conferences, seminars, and speeches have been dedicated to a related aspect of Quality than any other corporate topic. At times there is so much material related to Quality and Total Quality Management being introduced and recommended that it becomes overwhelming for a corporation to attempt to keep pace. Over the decades, Quality has transformed from a necessary evil into a marketing tool for achieving the competitive edge.

If properly designed and followed, the Quality system and effort will be the most effective tools in the prevention of potential product liability incidents in the field. In its best form, it is the single most comprehensive collection of documented practices and safeguards for ensuring that the corporation will produce a safe and reliable product for the marketplace. If needed at a later date, the records maintained will also become one of the most effective tools for proving in court that the manufacturing firm made every effort to ensure the production of a safe and reliable product.

THE EVOLUTIONARY PROCESS

Historically, during the '60s and into the early '80s, Quality systems and efforts had always been oriented toward product more than service. They focused on inspection, evaluation, and routine testing of the products, as well as on the control of other operations that could have an impact on the quality of the product. In the '60s and '70s, Quality efforts were focused more on Quality Control concepts, which translated into large departments of QC inspectors and backup personnel. Ensuring product quality during this era was truly in the hands of the Quality department itself, which therefore had to be large enough to handle the job.

Having only the Quality department be responsible for quality was a concept that eventually became self-defeating, especially when the rest of management was focused on sales and output. Those who could sell the most, followed by those who could produce the most, were immediately recognized and appreciated. Quality became nothing more than the bottleneck that seemed to keep the company from pushing out product even faster.

It took the Japanese Quality revolution of the '80s to finally show the American marketplace just how cheap and undependable our products really were. The Japanese, who themselves had previously had a reputation for producing cheap and unreliable products, had spent a few decades working diligently on improving their capabilities and building a reputation for quality products, making it the sole objective of the entire country. They ensured that every employee was totally involved and committed to producing the highest degree of quality. So, when they launched their invasion on the American marketplace in the early '80s, they took us by storm. And it was hard for Americans to remain loyal to domestic manufacturers and products when they knew they were paying more for our products and having more problems with them. This successful invasion significantly crippled corporate America and taught them a valuable lesson.

The lesson that came out of it was that corporate leadership had to shift its focus away from sales and output and concentrate on quality. In essence, quality would create sales, which also translated into increased production. Not attaining quality meant the end of the organization. And many companies that couldn't make the change vanished. So now, the Quality department and effort, which till then seemed like a lost voice in the woods, gained the attention it always deserved. The programs and efforts that these department heads were constantly fighting for were now being complied with and respected. New ideas and concepts were being introduced at a rapid pace. The methods that were so successful for the Japanese were being sought out by leading corporations across the country.

As the individual corporations began to develop knowledge and expertise in these new principles and concepts in quality that led to their own success, they began teaching others, and many of their staff members became consultants and prophets of what needed to be done by other companies that wanted to follow in their footsteps. For the decade that followed, those prophets and their programs grew as fast and as numerous as the products that were being made. The question soon became which philosophies were right and which were wrong. The programs and efforts that were generic up to this point soon became individual philosophies and corporate benchmarks and names began to be attached to them. No longer were they TQC or TQA programs; now they were the philosophies of Quality gurus, the secrets to success by major leading corporations, and continuous revolutionary breakthroughs by various authors.

Recognizing all this newfound enthusiasm and wanting to be a part of it were those in government, who set up their own principles and guidelines for the Quality program and effort. And they gained even more attention because they offered an award (named for Malcolm Baldrige) for the corporation whose Quality efforts they considered best. The corporations that pursued the award were not necessarily truly interested in ensuring Quality; I speculate that their enthusiasm was more along the lines of the determination of a CEO to win the Oscar of the Quality world.

Meanwhile, overseas (the Atlantic, to be more specific), industry wasn't so bent on ego trips, but they were still faced with the necessity of having to improve quality and set some type of standard for what was expected. So, as the United States became one large Hollywood, the Europeans went to work on a concept that they could all be in agreement with, and begin to follow. This was the ISO (International Standards Organization) set of guidelines and principles, which, to a Quality professional with any amount of years under his belt, significantly resembled the principles once established in the TQA programs. In other words, it dealt primarily with the product, and any element that could have an impact on the quality of the product or service provided, and didn't really incorporate all the wondrous aspects of TQM, such as teams and team leaders, employee participation and empowerment, and internal customer satisfaction measurements.

Since the European community was becoming so united in the development and backing of this new set of standards, and many American companies wanted to sell their products and services to European countries (not to mention the fact that the Quality movement in America was in a state of chaos anyway), many companies changed direction and pursued this single train of thought, which would eventually surpass all other concepts.

THE LEGAL WORLD TAKES NOTICE

With so much being said and written about quality and Quality programs in the media, it is no wonder that it began to be a household word. This Quality revolution captured the attention not only of every business professional across the country but of legal professionals as well, and they used it to their advantage. They never became experts or even rank amateurs on the topic, but they recognized how important and popular it was for a manufacturer to have a comprehensive Quality program in place, as well as how significant a tool it could be in prosecuting a manufacturer that didn't have much of a program in place, if any, and was being implicated for manufacturing a defective product.

Attorneys recognized that such corporate negligence in not having a comprehensive Quality program projected an image to the court of a company that does not make reasonable state-of-the-art efforts to ensure that they are producing a safe and reliable product, nor does it project an image of a concerned and caring company. This lack of positive attributes will carry a lot of weight with a jury that is deciding the manufacturer's ultimate fate. Without many of the basic Quality program elements in place, the corporation will be at a significant loss to prove that it took the necessary steps and safeguards in the design and production of the product in question, primarily because it is normally the Quality system itself that requires the documentation that will then provide the evidence.

The prosecuting attorneys realize all this and commonly begin to pursue it in their first set of interrogatories or requests for documents from the defendant. An interrogatory is a long list of questions, regarding the product itself as well as the operation, which is sent through the court system to the company that is the defendant in the case and must be answered within a designated time frame by a representative of that company. It might include questions regarding the inspection of the product while it was being produced. It might also call for the respondent to supply copies of documentation describing the reliability testing to which the product was subjected and/or documentation of any Design Reviews held. Or copies of the Quality manual, so they can determine from it what documents they want to ask for. This is all part of what is known as the discovery process, and the recipient of the interrogatory is legally required to supply the information requested, under oath and under the penalties of the law.

So, if a company receives this interrogatory—and they inevitably will—but they don't have any Quality program that they can truthfully talk about and decide to fabricate answers implying that they do have such programs and efforts in place, they will ultimately find themselves in a lot of trouble once they are more thoroughly cross-examined in court. But at that point it will be too late to retract anything.

CRITICAL ELEMENTS OF THE QUALITY SYSTEM

A multitude of ways are being taught to go about developing a corporate Quality program. This book doesn't expound on how the system should be developed; it only describes what the system should include for the sake of preventing unsafe products and designs.

In all the recommendations that exist for proposed Quality system content, probably the best list of ingredients is incorporated in the ISO standards. The primary focus of TQM programs is the people, calling for total employee involvement in the management process and for attaining total customer satisfaction internally and externally. When it comes to preventing product liability, the focus needs to be purely on the product. It is a technical product focus, not a management style concern. A super TQM program, then, may not do anything for you in a technical product focus and analysis.

Below is an outline of what a manufacturer's Quality program and Quality manual should be built around. The corporate Quality effort should entail these features and contain procedures that describe how these elements are handled.

Administration
The designation of an administrative head of the Quality effort.
A documented system of procedures controlling each of the functions listed below.
Verification that the procedures are being followed.

Human Resources
Verification that employees on the job are qualified.
Verification that each employee received the necessary training to do the job.

Sales and Marketing
Product promotion—truth in advertising.
Product warranty statements and policies.
A process for review of customer needs and use of product.
A review of customer purchase order contractual agreements.

Product Development
Qualification criteria for the product design.
New Product Introduction.
Formal Design Review.
Review of needed warnings and instructions.
Reliability testing of concepts, materials and processes being used.
Blueprint approval.
Control of design and specification changes.
First-piece approvals.

Supplier/Subcontract Quality Control
Control of selection and qualification.
Communication of standards and specifications.
Monitoring and documentation ensuring compliance to standards and specifications.

Control of purchase orders.
Terms of agreement, subrogation, indemnification.

Process Control
Documented processes to prevent variation.
Control of changes to documented processes.
Verification that documented processes were followed.
Product identification or traceability.
Documentation of in-process as well as final inspection and testing.
Identification and control of nonconforming product.
Equipment calibration programs.

Shipping and Handling
Packaging development.
Packaging instructions and printing.

Customer Service
Receiving and documenting customer complaints.
Summary and review of complaints.
Procedure for recalling products.

Quality Information and Feedback
Statistical analysis programs and efforts.
Retention of records.

DEFINITION OF ELEMENTS

An administrative head of Quality: It's fine to say that quality is everyone's job, or that Quality begins with the CEO. But once you get past all the rhetoric, you need someone to administer the program, ensure that it is being followed, and supply the necessary guidance when necessary. Serious dedication to Quality programs doesn't just happen because of some executive edict. In the day-to-day life of any corporation, production and sales pressures will always command primary attention, and the focus and concern for following the system and monitoring the output become secondary. So the program needs an administrator who ensures that they stay on track and is removed from the other pressures of the organization. This is a dedicated function, not an individual given the responsibility of wearing two or three hats.

A documented system of procedures controlling the functions: To ensure consistency on a continuous basis in everything the corporation does that could affect the quality of the products or services provided, they need to compile the procedures that have to be followed for the entire process; this documentation will then be included in the company's Quality manual. Each line of a Quality procedure should answer the question of *who* has to do *what* and by *when*, if it applies (see Figure 1). The format and style of the procedures themselves aren't important as long as the entire chain of events is documented in flow-chart form (see Figure 2), agreed upon, and signed off on by the parties who have to carry out the function. This means that the procedures shouldn't be dictated by others, nor can they be some generic mail-order set of procedures that are so ambiguous in nature that they can be interpreted in half a dozen different ways. They need to be written and bought into by those who have to carry out the responsibilities.

In Figure 3 we see a typical example of a poorly written procedure. Although brief for this example, the procedure demonstrates several problems commonly found in many organizations' systems. To begin, the procedure never states who has the responsibility or ability to carry out any of the elements. For instance, there is no mention as to *who* can initiate an ECR form. It doesn't identify who needs to approve an ECR form, only that it apparently is approved by someone. It fails to specify whom the form is forwarded to. Would it be the specific engineer? The Director of Engineering? The secretary? It doesn't identify who has the responsibility for distributing the newly revised drawing, or the spe-

GE GOODDEN ENTERPRISES	QUALITY ASSURANCE PROCEDURE CORRECTIVE ACTION	NUMBER 101 EFFECTIVE DATE 1/14/95 REVISION DATE PAGE 1 of 3 Total Quality System

1.0 PURPOSE

1.1 To describe the procedure to follow once deviations from established Procedures are detected.

2.0 SCOPE

2.1 Pertains to all Quality Procedures in effect throughout the organization.

3.0 PROCEDURE

3.1 Any employee that becomes aware of the fact that an established Quality Procedure is being deviated from, has the responsibility of reporting the deviation.

3.2 The employee is responsible for communicating the deviation to the Director of Quality.

3.3 The Director of Quality Assurance will investigate the situation and confirm the fact that there is a deviation.

3.4 If there is a deviation, the Director of Quality Assurance will initiate a *Corrective Action Notice* to the Department Head responsible.

3.5 The responsible Department Head will investigate the incident with one working day and implement the necessary corrective action.

3.6 The responsible Department Head will document their response on the *Corrective Action Notice* and return it to the Director of Quality Assurance within two working days.

3.7 The Director of Quality Assurance will perform a follow up audit to ensure compliance to established Procedure.

3.8 The Director of Quality Assurance will file the completed document in the Quality records.

AUDITOR:__________ DATE:	AUDIT	IN COMPLIANCE OUT OF COMPLIANCE

Figure 1 Example of a typical Quality procedure and form.

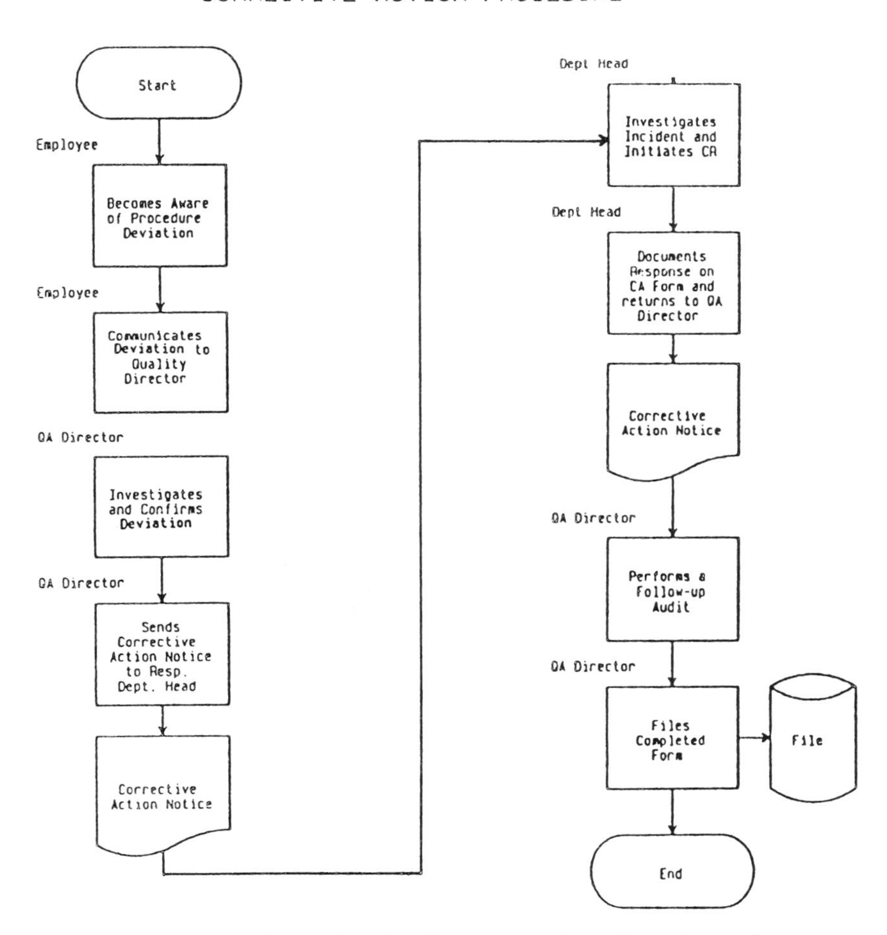

Figure 2 Example of the Quality procedure flowcharted.

GE GOODDEN ENTERPRISES	QUALITY ASSURANCE PROCEDURE CONTROL OF ENGINEERING CHANGES	NUMBER EFFECTIVE DATE REVISION DATE PAGE Total Quality System

1.0 PURPOSE

1.1 To control the process by which changes to existing engineering specifications and drawings are carried out.

2.0 PROCEDURE

2.1 No changes to engineering specifications or drawings will be implemented without the initiation of an Engineering Change Request (ECR) form.

2.2 The ECR form must be fully completed, stating the nature and reason for the change, prior to being submitted.

2.3 Once the ECR has been approved, the Engineer assigned to the product will make the changes requested.

2.4 The newly revised drawings will then be distributed to the various departments and individuals.

2.5 The ECR will be retained by the Engineering department for future reference.

AUDITOR: DATE:	AUDIT		IN COMPLIANCE OUT OF COMPLIANCE

Figure 3 Example of a poorly written procedure.

cific departments that it should be distributed to. This is somewhat subjective. It fails to state what those individuals or departments should do with their previous copies of the drawing. And it fails to identify who in the engineering department has the responsibility for retaining the ECR and for what duration of time. It is basically a list of statements.

As shown in the upper right corner of Figure 1, Quality procedures should have title boxes that show the procedure number, the date it was originated, and the date of the last revision. This would enable any owner of a manual to instantly know whether the procedure he is working from is current, merely by comparing the revision date to that of the controlled master, retained by the Quality administrator. The individual procedures should be revised and redistributed whenever they are found not to be accurate.

These continual revisions make it impractical to publish Quality manuals by professionally printing them. That kind of polished publication is the result of executives wanting to impress others with the professional appearance of their manuals, when in fact they are effectively showing that they are not on top of their Quality program, that it does not change with the organization, and/or that it is so ambiguously written that it is meaningless. Quality procedures are working documents, which means that they change with the organization as the organization seeks continuous improvement.

Verification that the procedures are being followed: This simply means routine audits of the Quality system. An organization would want to be able to prove not only that they had the necessary procedures in place to prevent potential deviations and errors but also that they knew the procedures were being followed. Audits of the Quality system would prove this point, provided that the audits are taken in a timely manner.

Verification that employees on the job are qualified: For every position in an organization there should be a job description detailing the responsibilities of the position and the skills and background required to be qualified for it. Comparing this job description with an employee's own résumé, along with the record of any additional training received, will show whether an individual is in fact qualified for a position.

Verification that each employee receives the necessary training: Employees at any level should receive ongoing training or their output and abilities will cease to grow. All training should be documented in some central file or computer database so that it can be reviewed when necessary.

Product promotion—truth in advertising: For those companies that advertise and market their products, it is important to review these ad campaigns and ensure that the messages being communicated aren't false or misleading, or that they could lead to an action that could create an unsafe condition.

Product warranty statements and policies: The warranty that a company offers on its products and services should be reviewed and updated on a regular basis. It should be clear to an individual purchasing the company's product what the exact warranty is that pertains to the product, along with any special notices of exclusion—especially those concerning conditions that would logically damage the product or render it unsafe to operate.

Review of customer needs and use of product: Prior to the design and manufacture of any product, the corporation must understand what the customer needs this product for, along with how it will probably be used. To prevent the potential for accidents that will likely lead to product liability, one must anticipate how the product is likely to be used by those who will purchase it, and design it accordingly.

Review of customer purchase order contractual agreements: Most purchase orders contain a substantial amount of fine print, which many times goes unread until some disaster transpires. In some cases, companies will supplement the fine print on the purchase order with other contractual clauses that will be forwarded to Sales. It is quite common for sales departments to ignore such contractual agreements and statements, feeling that it is just standard verbiage that one must accept in order to do business, and that the corporation need be concerned only with the warranties the company itself expresses or passes on to the customer. What may happen here, however, is that the customer can list what the supplier will be expected to do with regard to warranty concerns while the supplier feels that their own written warranty fully describes what they will do. In essence, you can end up with two contractual agreements that can be legally debated.

To guard against this, it is quite common for the fine print to include elements of indemnification, or subrogation, in the event that the product sold should ever become associated with, or be the alleged cause of, a product liability action. It is therefore necessary to ensure that these agreements and contracts are reviewed regularly by the manufacturer, in order to be aware of the customer's expectations and other legally binding clauses, especially as they may relate to instances of product liability.

Qualification criteria for the product design: The producer of any product should always be in a position to present the backup statistical information, formulas, regulative standards, and calculations that qualify the product design as being safe and reliable. To ensure the quality and reliability of any newly proposed product design, engineers have to base their designs and specifications on tested and proven concepts. In the face of a liability suit, the manufacturer will find that engineering a product based merely on theory or assumptions can be devastating to their defense and more easily challenged.

New Product Introduction: This procedure is probably one of the longest and most difficult to write, primarily because of all the departments involved in this introductory process. This procedure explains how the Product Introduction Team, or PIT, is established, and what its role will be in tracking week by week all the events in developing and introducing a new product into production. This will include the schedule of actions that must be taken to ensure that production begins when scheduled and that the customer gets his product when promised.

Formal Design Review: The Design Review is a meeting or series of meetings dedicated to reviewing the technical aspects of how the product is to be built, along with everything else that pertains to the product, by those members of management who have the technical knowledge and abilities to do so. The timing for this review is normally identified in the New Product Introduction procedure, but it may not include all the same people. This is one of the most critical elements in the Quality program, and another procedure that is often hard to write. One of the primary reasons that the procedure is difficult to write and implement is that it calls for several other departments within the organization that possess this technical ability to participate in critiquing of the proposed design, which engineering groups are commonly quite defensive about. For further details on the requirements for the Design Review, please refer to Chapter 6.

Review of needed warnings and instructions: Lack of necessary warnings and proper instructions is one of the top two complaints cited in product liability actions. The corporation needs not only to address this area of concern but to do so with more than just the engineer assigned to the project. This is one of the primary roles of a product safety team (which will be explained in following chapters), or at least a team composed of representatives from different disciplines that can look at this very objectively. The procedure must describe how this will actually be handled.

Reliability testing of concepts, materials, and processes: For a company to be innovative and remain competitive, it must constantly pursue new products and ideas. It is important, however, that these new concepts not be introduced prior to being tested and their safety and reliability verified. Proceeding with the use of such products and processes based only on supplier assurances is a major gamble by the final product manufacturer, and in the face of disaster he will likely find himself standing out there alone. The procedure for how the testing will be accomplished needs to be fully documented, along with what the test parameters should be.

Document approval: A procedure should be developed that clearly states how proposed engineering blueprints and specifications should be reviewed and approved.

Control of design changes: Once all the necessary departments have approved a product design from their own perspectives during the Design Review, it becomes important that no changes be made to the design without the knowledge and consent of these same departments. It is difficult for any one department to appreciate the impact that a change could have on everyone else; therefore it is important that such possibilities be guarded against. It becomes easy for the engineer, especially, to just go ahead and make various changes without anyone else's knowledge, out of a sense of ownership, but once these prints have been released they technically are no longer the sole possession of the engineer.

First-piece approvals: This is another critical first step in any product or component's life cycle. A good Quality system should identify who has the responsibility for checking the first piece of any product produced, to ensure that it complies with all standards and specifications that apply. It is also important that this approving party not be under the duress of various production pressures but be in a neutral position. The procedure needs to clearly identify how this is accomplished and recorded for future reference.

Selection and qualification of suppliers and subcontractors: A corporation that is concerned about the quality and reliability of the products they produce also needs to be concerned about what they are likely to be supplied. It is therefore imperative that suppliers and subcontractors be qualified prior to being selected. This is normally handled through on-site evaluations by the Quality department as well as by Purchasing. The objective is to avoid being influenced by an impressive sales effort of a potential supplier, by ensuring that all the mechanics are in place in a sound and reliable Quality system and effort.

Communication of standards and specifications: To ensure that a supplier or subcontractor will supply the needed product to the standards and specifications desired, it is essential that the corporation communicate said specifications to Purchasing, which then forwards them to the specific supplier. It is quite common to find that all the necessary information isn't being furnished to the outside supplier or that any changes made later aren't communicated. And, typically, the only contact within the organization that the supplier would complain to is the very department that was negligent: Purchasing. Therefore it is necessary for policies and procedures to be developed, and audited for compliance by a third party, Quality.

Receiving inspection or other documentation ensuring compliance to standards: There must be documented evidence that the material received through purchasing complied with the applicable standards and specifications. This can be accomplished through receiving inspection, or by having the supplier submit with each shipment certification records ensuring compliance. These records need to be traceable back to the product, and retained for a designated period of time.

Control of purchase orders: The PO is a vital link between the organization and its suppliers. In many cases, it is also the only legal contract binding the two companies. How the purchase orders are originated, what they should include as specification, what contractual agreements should be spelled out in them, and what should be sent along with them must be reviewed and written into procedure.

Terms of agreement as specified on POs, subrogation, indemnification: It is surprising how many companies don't pay any attention to the fine print in their own purchase order until they become involved in a product liability action—companies that fully expect to be indemnified by their suppliers but never said as much on the PO, or that expect subrogation in the event of an incident but have no such requirement stated anywhere. And, regardless of customer good will, it is the corporation's insurance company that will be paying the bill in the event of a product liability action. So, unless there is a contractual agreement for the supplier to indemnify its customers, the insurance company is not likely to honor a customer good will request. Suppliers and customers need to pay special attention to what each other is stipulating.

Documented process instructions: Critical processes in manufacturing must be fully documented and the documentaion available to the operator performing the specific function. It is recognized that new or transferred employees can enter the operation at any time; without the ability to refer to some type of process manual, the potential for variation increases significantly.

Control of changes to documented processes: Process instructions are the equivalent of blueprints, and therefore it is just as important that they not be changed without the knowledge and consent of other parties that may be affected. Procedures must identify how processes are developed, approved, and revised.

Verification that documented processes were followed: In addition to being able to prove that the important processes were developed and put into place, one also needs to be able to prove in some manner that they were followed. During a trial, proof that the critical process in question was being complied with is valuable information. This could be accomplished with records of routine documented audits or possibly the operator's own verification that the process was being complied with through the use of hourly checks or some other effective technique.

Product identification or traceability: Especially in situations where the product is made in high volume, and maybe for days on end, there needs to be a way to trace products to specific dates of manufacture or lots in order to identify other suspect product if any unit of product is found to be defective. The same holds true of components and materials going into the final product; they too need to be traceable in some manner as they move from department to department throughout the manufacturing plant, in the event that it is found that a specific lot or group is defective.

Documentation of in-process as well as final inspection and testing: Proof or verification of routine inspection and testing is a vital record in Quality Control. It is also an important piece of evidence for the defense in any product liability case. Although it may not be practical to document all the inspections a product may have received during manufacture, it is practical to document final checks, and especially reliability tests, that the product may have been subjected to.

Identification and control of nonconforming product: The Quality system should explain how nonconforming or defective product is handled once it has been recognized. How is the product identified, who determines disposition, what happens to the rest of the product in storage that may have the same defect? These issues need to be clearly addressed so that when such a situation does arise, not only is there a documented plan that goes into effect, but the situation is also documented to demonstrate that the entire situation was arrested.

Equipment calibration programs: The quality and reliability of a product are normally heavily dependent on the accuracy of the equipment that made it and its component parts. In essence, the quality of the end product can be only as good as the quality of the materials and processes that went into it. To ensure the consistent output of any process or piece of equipment requires that the same be calibrated routinely. It is essential that the frequency of this calibration be determined and incorporated into a procedure, and that the calibration check be documented and records retained.

Packaging development: Ensuring that a product will arrive at its destination in one piece (without being altered and conceivably transformed to create a defective condition), as well as be practical for the end receiver to deal with, is all part of carton development. Most companies don't engineer or manufacture their own packaging. The packaging is typically ordered by Purchasing and designed by an engineer from the packaging supplier. Just like any other first-piece approval process, packaging should be reviewed, tested, and approved prior to production.

Packaging instructions and printing: In addition to reviewing the package design to ensure safety and reliability, it is equally important to review the content and presentation of the instructions that will accompany the product and any necessary cautionary statements to be printed on the carton.

Receiving and documenting customer complaints: If Customer Service receives a complaint from a customer about a product, how is it handled? How is the complaint communicated to other departments that need to know so they can make any necessary corrections to products and processes? How does a corporation recognize a potentially defective product line and react, or recognize developing trends regarding problems with products? Procedures that establish how this whole area of concern is to be handled and addressed should be developed to demonstrate that the organization effectively reacted to any known product problem.

Summary and review of complaints: To determine if a defective condition involving a product in the field is getting out of control or if a serious trend is developing that may require management action, customer complaints need to be summarized on a regular basis and analyzed by management. This documentation and summary of field complaints will also be effective in product liability defense if it can be shown that there wasn't a defective condition present in the many other products that were out in the field. From the other end of the spectrum, recognizing the growing risk of a certain situation or condition can aid the organization in determining that a recall is necessary.

Procedure for recalling products: Every corporation needs to have a procedure in place that clearly outlines the recall process. Who makes the determination to recall a product? How is it handled? What are the agreed-upon guidelines for determining that a recall is necessary? It is quite common for a corporation not to know how they would handle such a situation if it has never arisen before, or only a few times. Ironically, this is one of the primary reasons that a plan should be developed. If a corporation is thrown into an immediate field disaster, or is at the brink of a potential one, it could easily become a chaotic event if such a master plan had not been developed back when everyone had a clear head.

Statistical analysis programs and efforts: For a corporation to accurately determine how well it is doing in all areas of production and customer satisfaction, it needs to statistically monitor the key performance indicators that it considers significant. These would logically include: scrap, rework, the number of returns and the reasons, the number of defects in any product or product line, process variation, number of customer complaints and the reasons, level of customer satisfaction, and any other key performance indicator that it recognizes. Without this analysis, companies may perceive that they are improving or that they know the root causes of problems, but such speculation won't be based on fact. In a high percentage of situations, it will be found that perceived notions were 180 degrees away from actual fact.

Records retention: Lastly, it is extremely important that the corporation indicate which records need to be retained, by whom, and for how long. Records are the most effective tool in defense. The lack of any records is the most destructive omission a manufacturer can make. Various departments won't recognize the importance of keeping many records once a transaction has been complete for a few months or possibly years, but they're not as likely to challenge a written procedure that states that a specific record must be maintained for, say, 5 years.

SUMMARY

This, then, is a minimum review of what a Quality system would consist of. It is primarily a list of procedures and programs that will significantly help reduce the potential for a product liability action to begin with, but it will also help defend the corporation if an action should develop.

A company may have a very elaborate Quality system in place, but it is important to ensure that it addresses the elements most associated with product quality and the technical product, as compared to the numerous systems and elements normally contained in TQM that are likely to be centered on administration and office processing dynamics. In recent years, corporations seem to have begun to feel that product quality is secondary to service and office systems quality. They need to have equal concerns in these areas, and even when the quality level of the product is not an issue with the external customer, retaining adequate product quality documentation could instantly become the issue. Although we have identified approximately 35 key programs and procedures, it is not unlikely for the entire system, once established, to consist of 50–75 programs and procedures. Although such systems and procedures represent bureaucracy to some degree, they should not be perceived as being *overly* bureaucratic. Every organization has to establish procedures. The lack of procedures breeds chaos. The key to not becoming overly bureaucratic is to keep the procedures simple. Quality procedures need not be complex. They can be as simplistic or complex as anyone desires. They just need to be there and be followed.

5

The Internal Expert and the Review Board

INTERNAL EXPERT

Choosing the Internal Expert

As the corporation enters this new field of endeavor, its first major objective is to select the right internal employee to head up the function. Unlike some selections that tend to capitalize on leadership qualities and professional appeal or others that purely favor technical knowledge and abilities, this selection must be based on *all* of those.

The right person to manage this function and role will not logically be found in Finance or Human Resources. It will not be a person who has been involved only in office management or similar corporate roles. It will not be someone hired off the street who lacks any knowledge of the corporation's products, let alone the technical aspects of engineering and manufacturing. And it does not need to be someone who has a legal background or degree. Although the legal system seems foreign to most people outside of it, it can be learned—the experience this individual will gain through routine involvement in litigation will give him or her all the on-the-job-training one needs to succeed at this.

From the opposite perspective, a candidate for this role should not be a plant supervisor, an engineer of any sort, or anyone else in a middle-management position. And yet the individual has to have the general knowledge of all these positions as it relates to the manufacturing processes and the technical aspects of the product.

The individual selected to head up this function must be a self-starter, capable of intelligently discussing any aspect of technical product or process, and have the administrative abilities to drive an entire program and effort from scratch. The individual should be able to leave on a moment's notice and fly to other parts of the country to meet with attorneys and insurance company representatives and represent the corporation's best interests. The individual needs to be a technical ambassador for the company and remain calm under pressure.

The person should already be reporting directly to the president or CEO of the corporation, and be able to command the necessary support and respect of everyone throughout the organization. The individual selected needs to be able to effectively communicate where the corporation was negligent or remiss in a situation that led to liability, without being offensive to any other departments or defensive of his own. The ideal candidate will currently be in a neutral position in relation to the design, engineering, and manufacture of the product, and therefore have the effective capability of fairly judging all three.

The most logical choice for heading this area of responsibility, in lieu of anyone being given the role full-time, is the head of Quality Assurance or Product Reliability. Typically, this individual already has this technical knowledge and administrative ability, and in all probability was instrumental in developing the company's other preventive programs and efforts.

Position Title

The first issue that executive management will have to address is the title itself (unless, of course, the existing title isn't going to change). This is somewhat difficult, because there are many possibilities to choose from. If the selected individual was already the Quality Director, there may not be any reason to change the title. It would be looked at as just an added responsibility. Management could expand on the current title and change it to Director of Quality and Product Liability. If executive management is hesitant about using the term *liability* because they feel it may imply that the organization has problems in this area, they could make the title Director of Quality and Product Reliability, on the theory that a liability incident logically stems from an unreliable product so *reliability* could be used in place of *liability*. Another choice may be Director of Quality and Legal Affairs, which could relate to anything in the legal area and doesn't imply anything negative.

If the position is being created for this primary role, the title could be Director of Product Reliability and Liabilities, or Director of Product Reliability. A common and popular title is Corporate Risk Manager, or just Risk Manager. Another choice is Product Liabilities Expert. The important aspect is that the title somewhat reflect the role, not only to the employees internally but to all the parties the individual will be dealing with on the outside, such as attorneys and insurance representatives. If you say that you're the Manager of Manufacturing Engineering to an attorney on the outside, and you handle product liability investigations and cases, the tie-in isn't immediately apparent and it would appear that you were assigned the role just for this one case.

The point that you would want to get across is that you are authorized to handle all these situations and have some expertise in the field. Otherwise, it becomes hard for the external parties to take you seriously and they tend to work around you, or maybe just copy you on correspondence after they have initiated various actions. For the sake of continuity only, with no implied recommendation that the title be selected, I use the title Product Liability Expert throughout this book in referring to this position.

Position Training

Once the role has been assigned, the next objective is for the individual to begin to gain the necessary basic training in some of the legal aspects of product liability. This book will serve as an excellent resource for this knowledge, because it not only is based on years of personal experience but also includes the best of many seminars and teachings as well. This can then be supplemented with seminars that deal with product liability offered by some of the larger universities and associations, as well as various legal books, periodicals, and relevant articles. A partial list of resources is offered below.

Seminars
The Defense Research Institute
750 N. Lake Shore Dr.
Chicago, IL 60611

University of Wisconsin—Madison
College of Engineering
432 N. Lake St.
Madison, WI 53706

Periodicals and Other Publications
For The Defense (periodical)
Products Liability: The Duty to Warn
Practical Guide to Controlling Products Liability Costs

Products Liability Pretrial Notebook
The Defense Research Institute
750 N. Lake Shore Dr.
Chicago, IL 60611

Product Safety & Liability Reporter
The Bureau of National Affairs
1231 12th St. N.W.
Washington, D.C. 20037

The Product Liability Expert, promoted from within, is not going to become an instant legal expert in this field, nor should anyone expect it. The knowledge gained will be acquired slowly, over months and years. Corporations that suddenly recognize the need for this capacity must remember that up to this point no one was focused on this area. Now that someone is, the company can surely afford to let the person learn the ropes, and in essence the whole company will learn simultaneously. (If the company becomes involved in potential liability suits in the near future, it will be assigned an attorney by the insurance company.) And once this individual gets involved in litigation and begins to work with outside attorneys, in addition to the training they will pursue on the outside, he or she will learn a lot about the legal process. The important thing is that it is now finally being addressed.

Once the role has been established and the candidate begins studying the subject, the importance of the new role must be communicated to other members of the organization. Although others may become involved in this area through team or committee, it is important for everyone to know that there is only one individual in charge. Situations of potential liability may have previously been brought to the attention of others such as Sales and Customer Service; now everything should be brought to the attention of the Product Liability Expert.

Other Considerations

If this new responsibility is delegated to the Quality administrator as recommended, the individual is already probably a member of the staff and the role will complement their other duties. This new role should not require a substantial amount of time and attention, unless, of course, the corporation is so overwhelmed with such cases that it now requires full-time attention. In that situation, a manager or director under the Quality administrator may be the secondary choice. Although it isn't intended to be a full-time position, years later the need may be recognized for it to become one. Or it may become a modified position that currently doesn't exist and include a few other related roles, such as Manager or Director of Product Reliability, which may include heading the efforts of a new testing laboratory or focus on new product development, analysis, and evaluation; or Director of Legal Affairs, which may include patent applications and searches; or other newly acquired titles with specific parameters of focus.

Although for the present this role isn't intended to be full-time, it will periodically be very time-consuming and at times demand immediate attention. All parties need to be aware of this. For instance, the Product Liability Expert may be working on the development of the system and other related programs when suddenly the corporation is served a summons dealing with an incident. The Liability Expert has to be able to drop whatever was being worked on and deal with this action. The corporation may receive an interrogatory (as explained in Chapter 4, a questionnaire that can be quite lengthy, requiring all types of information regarding the corporation and the product itself, and may inquire into several other areas of Marketing, Sales, Engineering, and Manufacturing) as well as a request for documents (which demands copies of every report, PO, and blueprint that has anything to do with the product), and require this information within a specified short time frame. This will be very time-consuming for the individual in this role, and the corporation needs to allow for this and anticipate it.

Likewise, the need for travel must be recognized. When a corporation first learns of an incident, it must make arrangments to get to the location as soon as possible to learn the particulars. Such investigations can't be scheduled for more convenient time frames. And if an outside meeting involves other attorneys or insurance representatives, arriving at a mutually agreeable date and time will be difficult enough without the Liability Expert's having to work around several other commitments. So it must be understood that the position has the potential to be pretty demanding at times, and the individual selected has to be available and flexible.

REVIEW BOARD

Forming the Board

Even though the Product Liability Expert will have the primary role of handling these new responsibilities, he shouldn't function alone. Just as a Quality Director can't ensure quality through that singular position, neither can the Product Liability Expert prevent liability without additional help. It takes the cooperation of others to make this program a success. This is why, along with Quality, most corporations also have Steering Committees or Quality Councils. It enables the Director to still be the expert in the field and guide the entire effort in the right direction, but it also allows for additional input, buy-in, and commitment.

This collaboration is needed in the liabilities area as well. Even though this one individual will be largely responsible for communicating and working with the outside world by handling investigations and other legal assignments, there is now a lot of work that needs to be done within the corporation, and the Product Liability Expert will need to draw on the assistance of others to be successful.

At this point the Product Liability Expert wants to form a team or review board to begin addressing the internal issues, such as reviewing product designs, labels and instructions, engineering and manufacturing procedures, employee education, and much more that will be dealt with in detail in the following chapters. I stipulate here that the Liability Expert should be the one to form this group, because he should feel comfortable in working with the other members. I would also recommend that this new group be referred to as a *board* as opposed to a *team*, which is already too commonplace, or a *committee*, which implies that it may be a project or assignment rather than the ongoing role that it is.

Because the function is so new, if the Board is formed by the CEO there may be a rivalry in the initial stages as to who should really head it up, and that should be avoided. If this position wasn't given to the Director of Quality, then that person would be an excellent choice to be a member of this new group. An alternative choice would be the head of Engineering, because there will be much to do in this area as well and, rather than take the chance of Engineering's becoming defensive and resist the improvements that are going to follow, it is better to represent them on the team. Other choices might include the head of Finance, because of the working relationship the individual probably already has with the insurance company and his familiarity with costs and premiums; the head of Manufacturing, for reasons similar to those for Engineering; and the heads of Research and Development and of Manufacturing Engineering, because of their logical knowledge of the product as well as the process. I recommend, however, that the group contain only two or three members in addition to the Product Liability Expert. In all probability, many individuals may want to be a part of this but will add little value. The membership should consist only of those who really have a working part in this function, such as Finance with the insurance carrier and/or Engineering with designs and specifications, but others should not be considered unless they equally have something to offer that a different member couldn't address.

Just like the Product Liability Expert, once the members have been selected they should think of a title for the group. Various possibilities include:

The Product Safety Team (as opposed to just Safety Team, which really should be oriented more toward Human Resources and involve other aspects of concern such as employee safety)
The Product Liability Management Team
The Product Liabilities Review Board
The Product Safety Review Board

I have used the term *team* frequently here because it is quite popular at present with most corporate management. The term also implies a certain camaraderie or common bond among its members which is naturally needed here. The term is being used too often, however, and it may suggest a junior-management group. Because of the serious nature of this concern, and once again for the sake of continuity, I use the name Product Liabilities Review Board throughout this book when referring to this group of executives.

One of the first objectives the Board should undertake is to help the corporation develop a Product Liability Policy Statement or, if the corporation prefers, a Product Reliability Policy Statement. Although the corporation may already have a Quality Policy Statement as part of its Quality program, it needs to be reviewed to determine if it demonstrates the commitment toward safe and reliable product design now being focused on by the company. An example of just such a commitment is shown in Figure 1. When signed by the CEO and incorporated into the Quality system, these measures then become policy and are administered and driven by the Review Board.

Our company is committed to making every conscience effort to ensure that the products we design, engineer, and manufacture will be safe and reliable to our customers and the end user in the application that the product was intended. We will accomplish this objective through the following means;

- Newly proposed product designs will be reviewed by the management team to ensure that they are safe and reliable and will function as required in their intended application.
- Newly proposed product designs will be reviewed to ensure that we as a manufacturer, or any outside selected subcontractor, have the ability and provisions in place to produce the product in a consistently reliable manner.
- We will ensure that all new products, components, materials and processes, are thoroughly tested and approved prior to their use in production.
- We will concentrate in design and engineering on "designing out" potential risks and hazards, and make adequate provision for warnings and instructions concerning any other "residual" hazards, as well as to warn against any "reasonably foreseeable" misuse of the product.
- We will ensure that new product designs and specifications comply with all regulations, standards and codes that apply.
- We will ensure through routine testing and inspection that the products being produced conform to the standards, specifications, and performance factors that apply.
- We will maintain documented records that prove that the above safeguards were addressed properly, and retain these records for a designated period.

These policy statements will serve as our guidelines in ensuring safety and reliability in the products we manufacture, and will be incorporated into the appropriate procedures of our Quality system. Compliance to these statements and guidelines will be the responsibility of every employee of our organization.

Assistance in the interpretation and application of these policy statements and guidelines will be given by the Corporate Product Liability Management Team.

____

President

Figure 1 Product Liability Policy Statement.

Routine Meetings

This three- or four-member Board would then work closely with the Product Liability Expert to begin to put all the elements in place to try to prevent incidents of product liability from surfacing in the first place. This process is very similar to the start of any Quality program. In the routine meetings, which should be held on a regular—perhaps monthly basis—the members should bring to light any new cases that have surfaced and updates on any existing cases, and discuss programs that need to be initiated to improve product safety and any training or awareness sessions that are planned.

Minutes of their meetings should be distributed to the executive staff, including the president, the CEO, the chairman, and anyone else who heads up the corporation. In this manner, even though several members of the executive staff may not be members of the Review Board, they remain aware of what is happening and being planned and have the opportunity to present any recommendations to the members. However, because of the serious nature of this subject matter, and especially the updates on new and existing cases, these minutes should be stamped "Confidential" and should not fall into the hands of the rank and file. Much of what will be reported regarding cases should not become common knowledge among the employees or it will become common knowledge outside the company as well. Second, the less the individual employees know regarding a product liability incident, the less the chances of it and them becoming a problem should they ever be brought into a trial as witnesses or deposed. This is sometimes hard for employees to understand, especially members of Sales who may have brought a specific situation to light in the first place and now expect updates on where it stands. Such updates should just be given orally by a member of the Board, and only when asked, and other unrelated situations shouldn't be discussed. What the rank and file needs to accept is that the situation is now in the hands of others who have the responsibility to handle it on behalf of the corporation, and technically neither the customer nor the sales representative need to remain involved.

The minutes or summaries of the Review Board's meetings should be brief, stating the facts as to the current status of each case but not the planned strategies. It must be remembered that any such document can be sought out at a future date by counsel for a plaintiff during the discovery process, so you wouldn't want the minutes to include defense strategies on new or existing cases, nor do you want others reading about members' discussion of an internal situation or problem that could put the corporation in a bad light (see Chapter 10, "Dangerous Documents"). For instance, if the Board recommends that a product incorporate certain printed warnings or instructions but the organization decides against it because of cost or fear of what a customer might think when they read them, the rejected proposal could prove fatal if discovered by a prosecuting attorney at a later date. If the Board wants to recommend an action regarding a certain concern that may be challenged by the executive group, the minutes could briefly reflect that the Board is working on the issue and readers can contact members for details.

I have described ways in which these meeting minutes could prove to be detrimental to the organization, but they can also be very beneficial for the defense in proving to the courts that the corporation was making every effort to address product safety and demonstrate that it is a concerned and caring company.

It is also important for the organization to institute the recommendations made by the Board, or the Board's effectiveness could be crippled and there might be other potentially negative repercussions. This Board must be given the responsibility for taking any action necessary to help prevent the possibility of future product liability, and these proposed actions should not be challenged by anyone other than the president. It is not like a Quality program where one seeks "buy-in" by the others. This is a dedicated group that has to decide on behalf of the organization what the company needs to do.

Even with an existing Quality system and program in place, corporations will find that this newly focused effort will bring about numerous improvements in the way they do business. The corporation will benefit from this Board and its specific focus, and from the new programs that they introduce that might otherwise have never come into play. In addition, this Board will bring together a few of the major disciplines (Quality, Engineering, Finance, etc.) to share in one common purpose, whereas they normally tend to operate independently. A department might learn that an action it initiated to address or simplify a process for itself has a potential reliability impact on another department. For instance, Engineering's focus on a product drawing may have been just to identify the components and specifications that pertain to the product and not make much reference to needed warning labels or associated instructions. Now they have more concern along these lines and recognize these other areas that also need to be addressed. Furthermore, as the Product Liability Expert gains additional knowledge and education in product liability law, and about how the product is being used on the outside, he will have a better forum for sharing this new knowledge and enabling the membership to react on the issues accordingly, such as by expanding the instructions and needed warnings.

When it comes to product liability law, not only do corporations typically lack very much specific knowledge on the subject but, more dangerously, many executives *think* they know a lot. And when the company begins to react based on speculative knowledge, they run a greater risk of creating problems than had they not reacted at all. When it comes to reports, memos, letters, and records, there is a lot that needs to be understood. The same is true of testing, warning labels, instructions, and the handling of other known hazards. There is much to be learned by the company in all these areas, and this is how they will begin to do it.

Multiplant Corporations

In a corporation that has multiple plant operations, the Product Liabilities Review Board at the corporate level may have to be complemented by similar teams at the plant level. In this scenario, the corporate Board may establish policy and guidelines as well as training and awareness programs, but they will need to depend on additional efforts at the plant level to ensure that the entire mission is carried out.

In this application, if the corporate group is called the Board, the plant team may be referred to as the Product Reliability Review Team. Like its corporate senior group, this junior organization would try to compose its membership from such departments as Quality, Engineering, Manufacturing Engineering, possibly the controller, and/or any other representatives of management the plant or corporation chooses.

The plant Product Reliability Review Team then monitors the activity of the operation at the plant level, involving any new products that it may be introducing as well as existing products that need to be brought up to standard, and brings these concerns along with their proposed action plans to the attention of their senior group. By having teams at both the corporate and local levels—two sets of eyes, as it were—the organization gains substantial assurance that their best interests will be addressed.

6

Design Reviews

The single most important event in a product's life cycle is the Design Review. This is when the technical product representatives of management get together as a group to critique the concept or the proposed design of a new product, and determine whether it will be safe and reliable. And yet, as important as this analysis is, as logical as it seems that a corporation would hold such a review, as economically opportune an event that it would appear . . . it doesn't happen. Or, in many cases, corporations just *think* that they're doing it.

Some companies have a small group of proprietary products that they continue to make in high volume and they really don't introduce new products very often. With this type of company, this whole concern is not an issue. But for the company that introduces several new products a year, or especially in a job shop atmosphere, this concept is crucial. These are the kinds of companies that this chapter pertains to.

In the Quality system, one of the most critical and probably most controversial procedures that is developed early in the program is the procedure for New Product Introduction. Normally this procedure describes the steps that the company will take in the design, development, and introduction of a new product it plans to produce. The procedure is usually so difficult to develop because of the number of departments involved in the process; it takes months for everyone to arrive at agreement. Logically this process will involve, at a minimum, Sales, Design, Engineering, Quality, Manufacturing Engineering, and Manufacturing, but it is also dependent on how the company is organized.

Within the context of this procedure will be the Design Review. There may even be more than one Design Review by a company, such as a:

Preliminary Design Review: An introductory review of the product that the company intends to develop and produce, which at this point may involve only pictures of a prototype or sample, or an artist's drawing of the proposed product.

Intermediate Design Review: Takes place once the proposed design and engineering have been completed, but before they are released. It would offer all the departments involved the opportunity to critique the proposed plans from all angles, and address any concerns they may have from their own perspectives.

Final Design Review: Takes place after the first prototype or first-piece sample of production has been completed in compliance with the standards and specifications developed, but prior to the launch of production. Once again, it offers the various disciplines the opportunity to evaluate the actual product, but now they have gained some working knowledge of the required fabrication and assembly. Once this review has transpired and all elements of concern addressed, production would start.

These reviews would be fully documented, and any concerns expressed would need to be addressed. Although the ideal is for all three reviews to take place, it is recognized that time constraints and the nature in which a product might come into being may not allow for all three reviews. But the corporation still needs to ensure that an effective Design Review took place.

NONFUNCTIONAL DESIGN REVIEWS

One of the immediate problems with a design review is that the attending parties may not be aware of the objective or of their roles. For example, Sales, or Sales and Engineering, displays a proposed new product and the attending department representatives view it in awe. They interpret the objective of the session as being to make them aware of the new product, to share the knowledge with them as if none of them had any immediate responsibilities, and that they can sit back and look forward toward its introduction. So all the departments and parties present walk out of the meeting and either feel positive about the product their company is going to produce or are skeptical about how it will perform or their ability to produce it. But little or nothing is said in the meeting itself. This was not a Design Review. It was a new-product awareness session.

Then there is the problem typically experienced with the designing engineer. When the product is being presented, the engineer dominates the meeting, takes ownership of the product, and dares anyone to challenge the proposed design. Basically, it's a "here's what we're going to do" scenario, and he is the uncontested expert. This also is not a Design Review.

Finally, there is the Design Review that consists of some of the Manufacturing management; the purchasing, scheduling, and materials departments; and the Account Manager. In this Design Review, the participants spend 5 minutes looking at the product and then immediately go into production schedules, planned production quantities, material leadtimes, supplier selection, and anything else that deals with preproduction planning. This is also not a Design Review. This is a scheduling and planning meeting.

THE ROLES OF THE PARTICIPANTS

A real Design Review is a critical first look at a proposed product or design, to determine if the product appears from every angle to be safe and reliable, whether it will truly satisfy the customer's needs, and whether the company has the ability to manufacture it in a consistent manner to the quality standards and engineering guidelines specified. There is no ownership of this product. Everyone needs to be critiquing this product and the required process for producing it on behalf of the corporation. In order to do this, the various departments must recognize their roles and responsibilities as they relate to their contribution in a Design Review. Here is a sample of those roles:

Product Engineering: This department's responsibility is to develop the blueprints and specifications for the product, based on engineering principles and guidelines, in conjunction with all known regulations, standards, and codes, in anticipation of the customer's intended use, and based on the engineer's own knowledge of the company's capabilities. The engineer is not expected to be an expert in the other manufacturing disciplines or areas of concern, but should be expected to have some working knowledge of them. The engineer is therefore bringing to this review a set of proposed product drawings and specifications based on this knowledge and experience, to be used as a proposed *starting point* for review and critique.

Sales and the Account Manager: Sales comes to this session representing the customer and the customer's expectations of the product. This could include various standards and specifications including color, materials, mechanical function, labels, and any other special notes that the group should be aware of.

Manufacturing Engineering: This department represents the company's manufacturing processing capabilities and manufacturing requirements, and normally consists of industrial engineering, methods engineering, and process engineering. Their purpose here is to understand the types of manufacturing processing that will be required, and how the plants and departments will handle the material flow, required processes, workplace design, and timing factors.

Manufacturing (departmental) management: These representatives should be the most knowledgeable of what their specific departments and equipment are capable of doing. They would come to this review not only to gain knowledge of the product and understand all the manufacturing requirements but also to review the proposed engineering specifications, tolerances, and standards and ensure that they have the capabilities to comply with them. The engineer may have a working knowledge of each of their abilities as a department, but these managers should be the experts along with the manufacturing engineers that work with them.

Quality and Product Reliability: Quality plays a number of roles within the organization. First, and as it primarily relates to a new aesthetic product in a job shop environment, Quality needs to secure any aesthetic standards that pertain to the product. Quality would also make note of certain dimensional standards being specified, and of what Sales communicates are any special customer requirements or specifications regarding the product, because these requirements then become the standards for evaluation.

From another perspective, Quality needs to be the conscience of the company. How reliable is this newly proposed design? What needs to be tested? What will be the processing concerns? What materials are they calling for, and what would the concerns be regarding those materials? What will need to be established regarding auditing programs or on-line testing? What are the concerns regarding packaging, labeling, and instructions? What special procedures or process instructions may this new product require? How will they ensure quality in everything they will be doing?

Materials management: This group is similar to some of the multifunctional departments listed above. It may consist of procurement, scheduling, material handling, and a few other functions. Their contribution to this review will be primarily from the perspective of the materials required and their availability. Who will they be using as suppliers? What might the problems be regarding the materials or the suppliers? Will there be any special handling or processing requirements?

Although this department is also concerned with scheduling and lead times, these concerns should not be discussed during this review. The purpose of this meeting is purely to critique the product and product design. The entire scheduling concern should happen during a different session, and most of these other departments may not even need to be there.

This, then, is a brief look at what the various departments would bring to the table during a typical Design Review. It isn't meant to be a session of "just watch and listen"; instead it needs to be a session that says "if there's anything wrong with what we are proposing to do, or if there's a better way to do it, now is your time to speak up, not later." This radically changes the perspectives that most companies have when they hold Design Reviews. Until now, the departments came to them to hear what was being said and see what was being proposed. Or they ended up being merely production or event scheduling meetings. But now each of the departments has specific responsibilities and reasons for being there, and they can't hold these roles passively.

It is common for many companies and their management to feel that they have Design Reviews, only to find that by the above definition they don't even come close. In most cases, they don't even have the right parties present to perform a real critique of a product design. A technical critique of the product would not be the strong point of production scheduling and purchasing departments. If the technical disciplines aren't present, then who's doing the critiquing?

IDENTIFYING OBSTACLES

Engineers typically hate Design Reviews. The ironic aspect is that, in many situations, they're the ones who have been given the responsibility to hold them. So they view it as an occasion for them to say (or at least feel): "Here's what we're going to do" or "Here's what I am proposing; come and chop it to pieces." I have been to countless Design Reviews held by, or taken over by, the head of Engineering. Engineers create something from nothing. An engineer typically has an ego. Egos become problems. If an engineer does a good job of designing a product, no one has any problems. If an engineer does a poor job, everyone has problems. If the Engineer tries to do it all himself, he creates the potential for problems. It wasn't intended to be this way.

Another real obstacle in getting a corporation to hold the necessary review is the argument that there isn't any time to do it. Companies view the challenge of competing in the global marketplace as meaning that the product has to go from concept to reality in a fraction of the original time that was normally allotted. If it used to take 3 years to introduce a new product, then we need to streamline that to 1.5 years. If it took 12 months, we need to now be able to do it in 6. And if it took 12 weeks, we need to turn the product around in 6 weeks, and keep getting faster at it.

There is no doubt that this is largely true; you can't just refuse to continuously improve. But just as it is important to shorten lead times, it is still just as important, or maybe even more important, to ensure safety and reliability. That same competitive world is not only demanding shorter lead times, but is also awarding continually higher financial settlements in product liability cases. But in the same manner that a management team figures out how to streamline the impossible, so too is there a logical way to ensure that a Design Review takes place without shortchanging its credibility or effectiveness. It always brings us back to the old saying, "We may not have had the time to do it right, but we'll always find the time to do it over."

THE PROPER METHOD

To begin the process, one of the primary candidates for scheduling a Design Review would be the Account Manager. This is the person who should be maintaining the schedule of events for launching the new product, including the time it takes to do the necessary engineering and lead times for ordering materials and for the development of tooling. Someone needs to put together this timeline and follow that things are really happening on schedule. And the Account Manager is one of the best people for this part, especially if the new product is for a specific account. If the product is proprietary for the company, it might be a Product Manager. So it really depends on the nature of the product and the organization. For the sake of simplicity, we will just use the term Project Manager.

In holding the *Preliminary Design Review*, the Project Manager features the principal player at that point, which may be the Account Executive or salesman, or the artist or designer. The Project Manager is in charge of the meeting and ensures that the necessary information is communicated, any elements of concern are documented, and any necessary follow-up takes place.

In the *Intermediate Design Review*, the Project Manager presents to the group the engineer with the proposed drawings. The engineer would lay out the drawings in front of the review team from the perspective that these are his thoughts are on how we would go about building this product, but encouraging everyone to take a good look at it and see if they can foresee any problems or any opportunities for improving on the design.

This is a revolutionary way for an engineer to present such work, with a radically different attitude toward it. The engineer is going from a possessive/defensive mode to an open-minded, we-all-own-the-product team concept. This is very positive and will yield the best possible results. In this atmosphere, the engineer doesn't take personally any of the constructive criticism offered, and no one presents it offensively. Everyone needs to be sensitive to the concerns of others, but at the same time they need to address any concerns they may have or recommend the best ideas for improvement that they are knowledgeable of, all on behalf of the company and the success of the product. If anything goes wrong from this point on, it is the team's fault for not addressing it properly, not the engineer's.

One approach to convincing the engineer to become part of this new process is to convey that he is being relieved of the sole burden of ensuring the credibility of the design; that responsibility will be divided among all these other management representatives. All the engineer is doing now is putting this collective train of thought down on a set of drawings and releasing them. Few managers wouldn't appreciate the relief of not having to make the sole decision on an issue and instead allowing a team to decide.

And in the *Final Design Review*, the Project Manager brings together for one last review all the same parties and department heads that have been involved in some fashion up to this point to take one last look at a sample of the product. At this point they have also gained the working knowledge of constructing one of the products and actually seeing how it functions. Naturally, however, other components are likely in process, so any changes here will be more expensive than they would have been had they been made at the previous review. Of course, from the other perspective, any changes here will be a lot less expensive than they would be during or after production.

It is easy to see how effective such meetings or reviews would be in identifying and catching potential problems. It also demonstrates the responsibilities each department now has in the review of this product. As previously stated, each of these sessions or meetings needs to be clearly documented, with all the expressed concerns being addressed. These minutes then become part of the historical record.

All the parties who were present during each of these reviews sign off on a Design Review form acknowledging that they carried out their required responsibility and are now approving the new product. As shown in Figure 1, the form may list the responsibilities of each department. Signing the form is a key step because otherwise anyone could attend a project meeting and later say that they never agreed to what was being presented. What we want here is to establish that everyone knew what their responsibilities were, acknowledged that they fulfilled them, and approved the product. If followed properly, this procedure is the best effort an organization could implement to ensure a high-quality and reliable product.

As previously stated, a corporation may find it difficult to find the time for all three kinds of meeting. In a job shop environment, they may even argue that because of the vol-

DESIGN REVIEW VERIFICATION

Product(s) Reviewed ______________________ Date__________

____

Customer__________________ Sales Order #__________________

Coordinator__________________ Location Held__________________

APPROVED BY

SALESMAN	_ACCOUNT MANAGER_
ENGINEER	_MANUFACTURING ENGINEER_
MANUFACTURING MGR _DEPT_	_MANUFACTURING MGR_ _DEPT_
MANUFACTURING MGR _DEPT_	_PRODUCTION CONTROL_
PURCHASING	_QUALITY_

RESPONSIBILITIES

SALES/ACCOUNT MANAGER - Ensure that all customer specifications, standards, and concerns are clearly communicated and that any standards or samples are passed along to the proper parties.
ENGINEERING - To present the proposed engineering concept ensuring that all parties understand the specifications and requirements, and solicit any ideas for improvement or identify any potential problem areas.
MANUFACTURING ENGINEERING - To note and plan the necessary processes needed to manufacture, ensure our process capabilities, along with recommending any ideas for process improvement.
MANUFACTURING - To understand the process and assembly steps required, and assure that we have the capabilities to produce the product to the standards specified.
PRODUCTION CONTROL - To ensure that the plant is capable of meeting the customers production requirements.
PURCHASING - To identify the materials required can be procured in the time allotted.
QUALITY - To ensure that all customer standards are obtained, and reliability concerns are addressed.

Figure 1 Design Review verification.

ume of new products being introduced, they can't find the time to hold even one of the meetings. But, technically speaking, the review doesn't really need to be a meeting. It would be the most recommended way, but for a corporation that has a real problem with holding any more meetings than it already does, there are alternatives. Basically, since the departments have documented responsibilities in this procedure—areas of concern that each of them is required to address when signing off on the form—each of them could technically review the product separately, note their concerns with what is being proposed, and/or sign off on the form at their own convenience. In this manner, all the responsible parties would eventually review and approve the product.

THE COSTS OF NOT HOLDING DESIGN REVIEWS

The Design Review stage is the least expensive one at which a company can react to a potential problem. In the past decade, numerous major corporations from around the world have told horror stories about missing this opportunity, because either they neglected to have them or they held them in haste. Based on these experiences, they have also shared information about the economic penalties they paid.

In one situation, within a couple of months after launching a brand new piece of office equipment across the country, a major corporation found that the product was defective—it was a problem that could have been caught in a Design Review had they held one effectively. The costs for correcting the defect at the various stages, and the actual cost of catching too late, were presented in this manner:

Defect identified and corrected during Design Review	$35
Defect caught after review but before part procurement	$177

Defect caught after procurement but before assembly	$368
Defect caught after assembly but before shipping	$17,000
Actual cost of correcting problem in field	**$590,000**

In this single example, one can see that the actual cost of not catching a defective condition at the $35 stage in the product's life cycle ended up being 16,500 times as much money. Unfortunately, it sometimes takes a catastrophic event such as this for a company to recognize the need for ensuring that these reviews take place; hopefully just the sharing of this type of knowledge could put the necessary fear into a company.

But this incident isn't unique. I knew a Quality Director for a large, high-volume electrical product manufacturer who experienced two situations within 5 years. A $3 and a $5 electronic component in two entirely different products caused the corporation over $200,000 each to correct, because the product was already in the field before the defect was identified. In one case, the defective condition could easily have been caught during either the Design Review or the routine reliability testing that should have been started shortly thereafter. In the other situation, the defective condition was in a supplied component and should have been caught in the supplier's own Design Review.

In all these examples, however, the companies were lucky that the defective condition resulted only in the mechanical failure of the product and not in a product liability incident. Take these same actual costs and throw on top of them a single $3 million cost because someone was severely injured or killed, and now see what you're looking at. So, as we evaluate this potential from this new perspective, we need to multiply the possible cost by many times the mere cost of recalling or repairing.

THE NEWEST MEMBER OF THE TEAM

Now we add to this Design Review our newest member of the team, the Product Liability Expert. Up to this point, we have reviewed and analyzed the product from all the angles and areas of our own expertise: process capability, manufacturability, standards and specs, and product reliability (from the perspective of its intended application), just to name a few. But now we have a player entering the picture who will be taking a critical look at this product from an entirely different perspective.

The Product Liability Expert will be evaluating the product with the following questions in mind:

Is it a reasonably safe product?
What could possibly go wrong with this product?
What might the end user do with this product that we are totally overlooking?
Are the dangers obvious?
Are safety devices absent from the design?
How could it possibly lend itself to property damage?
What kind of warning labels or instructions should we be thinking about including with this product?
What would attorneys call "reasonable foreseeable use" for the product?
How does it relate to the state of the art in the industry?
What legal complications could we envision by introducing this to the marketplace?
Are there any codes or regulations that this may be expected to comply with?
What could be some extreme applications for this product, climatic or environmental?
What types of unique tests should we immediately undertake to ensure that the product or its materials will prove to be reliable?
What similarities does this product have to others that may have historically led to problems in the field?

This is a new perspective of analysis. Although some of the areas of concern may to some extent have been raised or considered by other departments, it would not have been with the same focus and intensity with which it is now being addressed. Nor would it have been from the same level of awareness and eventual expertise.

Some of the elements may have previously been brought to light by Engineering themselves, but could have been compromised by Sales because of seemingly more important concerns regarding aesthetics, function, or competitive costing. But now the individual bringing these concerns to the table is fast becoming the in-house expert on the subject. And, as time goes on, the credibility of what the Product Liability Expert says or asks will grow with his experience.

With this addition to the staff, the organization should benefit substantially in its efforts to prevent defective conditions from surfacing in the first place. This adds to the Quality and Reliability efforts already logically in place, and brings in the potential liability aspects.

In the Intermediate Review, for instance, the group has the best opportunity for recognizing potential hazards in the new product and making every effort to have them "designed out." If the potential hazard cannot be eliminated, the question then becomes whether there is a feasible way to protect the end user against the known hazard, such as with guards, other types of protective devices, instructions, and warnings. The questions that might be asked in addressing this are: what alternative approaches are available; what are the cost comparisons between this design and the alternative approach; what are the risk levels associated with any design; what should we incorporate into the instruction sheets to make the hazards or dangers known and explain how to prevent them; what other labels should be added to the product to prevent accident or injury?

In determining the level of risk associated with a known hazard and the degree of reaction required, one needs to evaluate the end user's potential exposure to the risk, and act accordingly. For instance, if a potentially hazardous condition is identified, but the possibility of exposure is determined to be *very remote* or *improbable*, the organization may be less likely to develop any type of safeguard, or even to act at all.

If the possibility of exposure to a recognized hazard is considered *less likely* but *possible*, as opposed to *improbable* or *remote*, the organization may elect to react by incorporating warning labels and instructions. If the exposure to the recognized risk is considered *routine* or *on a regular basis*, the organization would logically decide to incorporate mechanical safeguards or other protective mechanisms, if in fact the hazard or danger cannot be designed out altogether.

A MAJOR LEGAL TARGET

Just like the Quality system itself, the Design Review is a major focal point in litigation. All the questions we just listed will be raised again in litigation. The efforts that the corporation expended to ensure that it was introducing a safe and reliable product, and that it was truly state of the art, will be analyzed in depth by counsel for the plaintiff and used to help their case with the jury, if the corporation was negligent in this area.

In conducting a thorough Design Review and considering whether this newly developed product is really state of the art, the team needs to fully understand the term. State of the art means, based on the latest innovations in the specific field of technology, what elements could feasibly have been incorporated into this product to ensure its safety and reliability. To say that the product complied with the industry custom or standard of construction does not necessarily mean that it is state of the art.

Many legal reference books on product liability specifically expound on this subject of conducting a thorough Design Review. So, when an attorney initiates an action against a manufacturing corporation, one of the first moves is to submit the interrogatory (legal questionnaire) and the request for documents, which will likely deal with all aspects of the product's introduction. The manufacturer, on the other hand, needs to be in a position to demonstrate that all the appropriate preplanning and evaluations did in fact take place, and be able to present the documents that support this. Of course, if this really hasn't been the practice, the corporation will be at a total loss to develop anything at that late date. This is why the procedure needs to be developed, and followed, as soon as possible.

In addition to the keeping of these records, there is also a major concern for what they contain. In the following chapters we discuss this in great detail, but the bottom line is that it could be just as damaging to have Design Review documentation as to not have it. Once again, this is where the new Product Liability Expert will assist the company and supply the necessary education.

7

Reliability Testing and Inspections Programs

It would almost seem that if one were to rank the importance of a manufacturing company's various efforts as they relate to product liability prevention, they would be: 1) the Quality system, 2) Design Reviews, and 3) reliability testing and routine inspection programs. What efforts did the company undertake to ensure the safety and reliability of the product prior to its introduction to the marketplace would be a very common question asked by the judicial system.

Although companies may perform various everyday inspection tests on the products they produce, many of which may take place right on the assembly lines by production personnel, these aren't necessarily what are referred to as reliability tests. Reliability tests are typically controlled lab tests, pushing the product to its extreme in many different areas. It might entail extreme temperature tests, high- and low-end voltage tests, dynamics tests, drop tests, and other types of controlled tests to ensure that the product will withstand what it may be subjected to.

Reliability tests as well as routine inspection tests play an important part in ensuring the quality of the product. *The quality of the final product can only be as good as the quality of the parts and processes that go into making it.* Maintaining a high level of quality throughout the manufacturing stages requires testing. The types of testing that will take place are normally decided by Quality, Reliability, Engineering, and possibly Manufacturing Engineering, even though the tests themselves may be carried out by regular production workers. The testing should prove that the products or components comply with the established standards, which may be company, industry, or regulatory standards.

PROCEDURES FOR TESTING

The manner in which a product is required to be tested should be well documented in the Quality system of procedures. The procedures should describe *who* is responsible for performing the test, *what* the actual test to be performed is, *when* the test is to be performed, *where* the results of the test are to be documented, and *how* the individual performing the test should handle situations in which the product fails. Key elements being stipulated here are that the testing requirements need to be documented in procedures, the company needs to ensure the procedures are being followed, failures need to be identified and communicated, and the results and dispositions need to be documented.

In the earliest stages of product development, before going forward with the proposed design, reliability tests must be performed on the products to ensure that the product will live up to its expectations. When this involves purchased components and materials intended to go into the final product, it is easy and common for companies to place this responsibility for reliability testing on the suppliers of the product's components or process materials, and then expect to hold them responsible if they experience field failure. Although to some degree the company protects its own financial interests by holding the supplier accountable for field failure, it isn't recommended, and there will likely be a shared liability if the failure results in other damage or injury. If a company's product fails, technically, so does the company. A customer with a defective new piece of machinery isn't relieved to know that the problem with his new machine wasn't the fault of the machine builder but was due to a faulty switch that some other vendor supplied—he's upset that the machine doesn't work. Only the machine manufacturer is relieved to know that the fault lies with someone else, and they maintain a false sense of security if they assume the customer will accept this as an explanation.

Companies need to recognize the necessity for testing the reliability of the components they use as well as the end products they produce, especially as they relate to potential product liability concerns and product safety. If their end product fails, they are just as much at fault as the suppliers of the components, and they will be equally prosecuted by the law. And in many cases, the supplier of a component has no idea how the manufacturer intends to use their product (since manufacturers often don't involve suppliers in design reviews, which can also be a mistake), and therefore the supplier may not even accept responsibility once the application is known.

It is common for the salesman of any component or material to say that it will live up to your expectations, in order to ensure the sale. In many situations, they may even believe it. But the manufacturer of the final product needs to make sure that the new item can withstand the application. One way to accomplish this is to request the backup data that support the claim and critique the methods used to obtain the data. The other way is for the manufacturer to test the component themselves.

PERFORMING RELIABILITY TESTS

The need for this information and assurance is what leads the Quality Assurance departments in major corporations to develop their own reliability testing labs. They realize that the only way they will be entirely certain that the product will function in a reliable manner is to prove it themselves. This also gives them the most technical knowledge of what works, what doesn't, and why. It is amazing how much can be learned by performing reliability and failure tests on products and materials. It also becomes the basis for fact, while all else is speculation.

Engineering designs a product based on theory and speculation. They're basically all they have to work with. Product reliability testing proves, or disproves, theory and speculation. The more management and engineering resources a company has, the more theory and speculation it has to offer. Corporations have to be in a position where they are working only from fact. Putting into the marketplace any product that has the slightest potential for injury or damage is a foolish gamble, much less putting thousands of the products out there.

There are probably three classifications of reliability tests: routine performance tests, extreme conditions tests, and foreseeable abuse tests. This last classification is an entirely new perspective that we're adding, but it will take the corporation well beyond the norm in helping to prevent incidence of product liability.

Routine Performance Tests

Routine performance tests are the basic tests that any company must have in place. These are the tests performed on the product on the assembly line or at some other manufacturing stage to ensure that the product appears to be functioning properly. Checking whether an electrical product is functioning properly, a motorized product is moving as expected, or even that the electronic commands on a programmable product are communicating properly would all be typical routine performance tests. The need for high-volume production should never overshadow the necessity for routine testing.

The Quality system by itself will normally describe the various routine checks and tests to be performed on each product. This is naturally dependent on how the system is written and what safeguards were established. The system would identify incoming inspection tests to be conducted on the component products received and other tests that would typically be performed on subassemblies and fabricated parts typically throughout the manufacturing process.

It must be recognized that even if a product was inherently safe by design, last-minute decisions on the part of production or materials management or other personnel to make changes to the product or its components or processes could render the product defective. This might be the result of changes or substitutions in materials or process modifications that may go unnoticed by the parties who would recognize the implications.

Even though such changes are logically controlled by Quality procedures, the backup system for identifying the defect or deviation could be the routine performance test. Normally when the right tests have been performed and a defective condition is detected, the ensuing investigation into its cause inevitably reveals that someone deviated from the standard, creating the failure. Without such routine or statistical performance testing, the deviation may not be identified until after much of the product has already entered the field. It may even be first found by the customer.

Just as with any other quality concern, when products are found to be defective during these routine performance tests the questions to be addressed are: What is the nature of the defect? What caused the defective condition to develop? What needs to be done to correct the defective product? How many other products could have the same defect? How are these types of defects going to be prevented in the future? The routine performance test is a good quality check of the products being produced. The type of test to be performed and the process for doing it should be well thought out by Quality and possibly Manufacturing Engineering and documented in the appropriate, readily accessible process instructions, and the results logged by the employee.

In some situations, the performance tests are not conducted on a daily basis but have been performed by the company's lab prior to production. It may be a controlled test of the product to ensure that it will live up to its expectations and withstand the application. This wouldn't necessarily be a test of extremes, but more of an initial reliability or life test. If a company that makes doorknobs has determined that the average doorknob will be turned six times a day, and they want to ensure that their product would last a minimum of 5 years, they could create a mechanical test that would turn and pull on the knob 10,950 times to simulate that life expectancy. Once they've proven the design, they may go into production and never repeat the test.

This type of routine performance testing is seen in many commercials, demonstrating, for example, that phone handles were dropped on their cradle so many times without becoming defective, car or truck doors were opened and closed thousands of times, keyboards were mechanically hit with a fake finger over and over, or the classic suitcase throw test. These are routine performance tests on the product to establish for a fact that the product will withstand the normal wear and tear it is likely to encounter.

Extreme Conditions Tests

If a company gets the necessary testing equipment, it can also conduct the extreme conditions tests, subjecting the product to the ultimate temperature, climatic exposure, voltage condition, dynamics abuse, or any other condition of extreme exposure that the product might possibly be subjected to. With this type of testing the company's personnel gain even further insight into how their products will perform once they are subjected to these extremes. For instance, if the company manufactures a product that is supposed to function and remain outdoors, under routine performance testing the product could be tested at 70°, or they may have the opportunity to test it outside during the winter at 15° or on the hottest day of the summer, which may have been at 102°. Although the testing conditions may appear to be at somewhat extreme levels, the fact that certain parts of the country can get as cold as –50° and as hot as 120° shows that they really weren't extreme tests.

But going beyond the routine performance tests is what the company really needs to do to know for sure how their product will perform under conditions that it really might be subjected to, even, for instance, a company that makes an indoor display product. If there is a chance that the product could also be placed in a window, that changes the situation considerably.

A decorative black plastic LED clock that just hangs on the wall will probably see only the kinds of temperatures that you would expect to find in an ambient room condition. Put the same product in a window with the sunlight beating down on it, and the clock temperature may go beyond 150° just from the exposure to ultraviolet radiation. How will the electronics in the clock perform at those temperatures? What will happen to the plastic box that contains the clock? Suppose the electronic components were rated only to 125°, which is very typical. You know they would surely fail at this higher temperature. But how would you ever have known if you ran tests only at room temperature? And what would failure mean? What will the electronics actually do when they fail? Will they just cease to work? Will they degrade and catch fire? These are very serious concerns and the corporation needs to know the possible consequences. Failure that means merely that the product will quit working is one thing; catastrophic failure is something else entirely.

Performing extreme conditions tests usually means purchasing special testing equipment. The apparatus for accurate accelerated weather testing or exposure to sunlight with true correlation normally costs around $40,000 to $60,000. Poor imitations could cost as little as $5,000, but if the simulated light output isn't relative to true sunlight, what is the sense of the test?

Environmental chambers that can take products down to –40 or –50° and hold them there can be expensive as well. Ovens are probably the least expensive testing equipment, but even a lab oven big enough to put an average product into could cost $12,000 to $20,000. And, depending on the nature of the products, a whole assortment of other testing apparatuses could be necessary. This would definitely take the testing off the shop floor and place it in a special room or laboratory. If handled by the Quality department, it also creates a new function, something along the lines of a Reliability Testing Engineer. The knowledge gained by conducting reliability testing at this level is invaluable. Now the corporation will gain insight into what actually happens to their products when exposed to extreme, but not improbable, conditions. As time goes on, the corporation is likely to acquire more and more equipment, thereby expanding the lab's capabilities. The knowledge and insight gained will be of substantial help to the corporation in the design of new products or the use of new materials.

Another aspect of this laboratory that could be very attractive to the company, especially for those whose products are regulated by such agencies as Underwriters, is that the lab could qualify as being UL certified. This would mean that the company could test their own products, submit the results to the independent agency, and gain their stamp of approval—minus the wait and some of the expense. To perform all the tests required for such agency approval requires that the company secure all the necessary testing equipment. The testing that an independent agency would conduct on a product would be relative to what the product would be subjected to anyway, so having the same equipment and performing the same tests is the eventual objective for the manufacturer.

If a corporation developing its own lab sees this as their ultimate goal, they should pay special attention to the types of equipment purchased. Agencies rely on data from pieces of equipment that have certain specific accuracies, such as lab ovens, electrical meters, and burn testing booths. So, before purchasing the equipment necessary to develop an extreme conditions lab, one should decide on the future possibilities of this certification and purchase equipment with that objective in mind; otherwise the company may end up purchasing much of the equipment twice: the first time for less expensive equipment, then eventually replacing it with the right equipment.

Even if the company doesn't aim to have a certified laboratory, the capability to perform certification or ASTM tests will save the company a lot of time and money by being able to conduct the necessary tests ahead of time and determining whether the product is adequately designed. If a company submits a newly designed product to an outside lab or the standards lab for testing and approval, such testing could take 6 or more weeks and cost a couple of thousand dollars. If the product fails, it all has to be repeated. Having this capacity in house means that the product submitted has already been tested and proven out, and therefore the chances of its failure would be minimal.

Foreseeable Abuse Testing

This represents a radical new dimension of reliability testing. It is a takeoff on extreme conditions testing, and its results may sometimes be disputed by the engineering functions or executive management, especially when the product fails.

But it goes beyond that level of relative testing, and hinges on the legal concept of *reasonably foreseeable misuse*.

As a company enters the field of product liability prevention and handling, and especially as they begin to experience incidents or become involved in actual cases, they will find that this third dimension of reliability testing isn't as impractical as it may first seem. Reasonably foreseeable misuse is what some management might classify as abuse and wouldn't deem worthy of any attention. After all, "Why was the user doing that in the first place with the product? They should have known better. It isn't our job to have to design this product for people that are going to use it in various abusive manners that it was never intended for. I designed the product to be used only in its intended fashion." But then an opposing attorney representing an injured client steps in and says, "I don't think what my client did with this product should have been so unexpected. I think the manufacturer should have fully foreseen that someone might do this with the product."

Such unanticipated use or abuse could additionally stem from the use of a promotional adjectives such as "the indestructible . . .," the "totally safe . . .," or other variations of extreme abilities that might encourage daring behavior. They might be marketing strategies used to promote the uniqueness of the product, without any consideration for the adverse potential. But some situations may not be the result of questionable abusive use, just the result of ordinary behavior that would leave questions as to whether the end user should have known better—something as simple as using the product outside when it was logically intended to only be used inside, resulting in electrocution because it got wet. If the product didn't have labels or instructions warning against such use, the manufacturer may be liable for the consequences. The courts would eventually decide on such cases based on what a *reasonably prudent* person might think about doing with the product.

In routine performance and extreme conditions tests, the product is tested in its original condition, even if the tests go to the extremes. But in foreseeable abuse testing, the product might be altered. For example, the grounding prong might be cut off of a three-prong plug, which is commonly done by people who lack three-hole outlets.

What would happen if someone stuck his fingers or arm into a motorized or rotating device? Is there a built-in clutch? Would the motor just stop turning? Or would the person lose his fingers or arm? The courts would say that if the manufacturer knew of the potential dangers and consequences, they had a responsibility to warn the user.

What happens if you create a shorted condition in an electrical product? Does it actually trip a breaker? Does it just pop and expire or does it arc and ignite? Can the user receive an electrical shock? These are some of the situations the Product Liability Expert would explore in trying to determine a product's safety factor and in gaining more knowledge of the product's capabilities and potential. For instance, if the Liability Expert becomes aware of a situation in which the electrical product did arc and catch fire because it allegedly shorted out or the outside situation allegedly resulted in electrical injury and he knows the product has that potential, then he also knows that the allegation may be true. If, on the other hand, the Liability Expert knows from experience that the internal ground fault interrupter or thermal protector would be triggered and prevent such possibilities, then he could question such an allegation.

Take a situation as simple as that of a company that manufactures a swing set for a public location. It is fully intended for children who are probably under 10 years of age and weigh under 70–100 pounds. But when a 230-pound adult swings on the set with his child, the chain breaks and the individual suffers permanent back injury and sues. The manufacturer claims that the swing was never intended for an adult to swing on. Counsel for the plaintiff states that there were no visible warnings against such use and, since the swing was in a public location, it should have been *foreseeable*. The manufacturer may know that they designed the swing for someone under 100 pounds, but they may in fact have had no idea how much it could really withstand before it broke or what element of the swing would break first.

The following example is a little more technical but gives a good demonstration of the three types of tests. Fluorescent lamps are commonly used in many types of lighting fixtures, including ceiling fixtures, suspended lamps, and outdoor signs. Fluorescent lamps are powered by a ballast—a transformer, in a sense—which increases the voltage and regulates the amperage. In an outdoor sign, there may be three 6-foot lamps powered by one large heavy ballast. All the lamps, the ballast, and the wiring are contained within the sign and protected from rain and snow, although the sign itself will be subjected to extreme temperature ranges. The ballast manufacturer makes thousands of these ballasts and ships them to end users all over the country for all types of applications. As part of the quality check, the ballast manufacturer conducts routine performance tests on a statistical portion of the units produced, by wiring up the ballasts to a bank of fluorescent lamps and monitoring to ensure that they are working. To carry this test further, the ballast manufacturer takes other sample units into a lab and wires them to a bank of lamps, and then places the test units in a chamber where the ambient temperature drops to –30° for, say, a 24-hour period. In another, simultaneous test, they take a different bank of lamps and a sample ballast and raise the ambient temperature to 125° degrees and monitor the ballast's performance. This would then be an extreme conditions test.

Even though an outdoor sign box isn't supposed to get any water inside it, the ballast manufacturer realizes that it will happen because the signs aren't always watertight. Water can drip in from bolts in the top of the cabinet, from screw holes around the face, especially where the screws are missing, and from other areas of the sign drum or cabinet. Even in a ceiling lighting fixture, where one would think that there is almost no chance that the ballast will ever see water, it is common to see dripping water because of leaking roofs or plumbing. So, even though the ballast manufacturer could claim that the ballast wasn't intended to be subjected to any moisture and isn't responsible for any consequences experienced, they still realize that it might happen.

If the ballast manufacturer decides to conduct its own foreseeable abuse test, and in this lab environment with a sample ballast wired to a bank of fluorescent lamps creates a device that will periodically drip water on the ballast, they can see exactly how their ballast will react. If the water ends up being drawn into the ballast, because of the internal heat and wicking phenomenon, they might see their ballast short out. If, based on this test, the manufacturer improves the design and thereby prevents the possibility of failure, they will substantially prevent the potential for product liability and ultimately benefit from these efforts. From this learning experience the manufacturer has also created a better product, which will soon be realized by the product's users and will likely result in increased sales.

In this last example we see not only a demonstration of all three types of testing in one product line but also that there may be substantial benefits to acknowledging potential *misuse* as opposed to taking a stand against it. Recognizing that a certain abuse is likely to happen and designing around the potential can improve the reputation of the company and its product, even though the product was never intended to be subjected to the abusive action.

A good example of this is the original telephone-company phone. The phone receiver was never intended to be bounced off the floor, but by designing that potential into the product they created a better product on the market, and benefited from it. When the phone-company monopoly was broken up and other companies were coming out with phones that you could purchase for a substantially lower price, it was common to find that the less expensive phones could not withstand such abuse, nor would they work as well in many other ways. People didn't deny, when they dropped their phones and they broke, that they had created the result through their own fault and abuse. But knowing that the phone-company phones were built to withstand such abuse led many of them to purchase phone-company phones, and it forced the other manufacturers to produce products that could withstand the abuse if they wanted to remain in the business.

If the extreme conditions or foreseeable abuse tests aren't conducted by the Product Liability Expert's own department, then he needs to maintain a close working relationship with those who are doing it. Such working knowledge cannot be gained through any other means, especially as it relates to the company's own products. So, to acquire the knowledge necessary for this function means that one has to either work closely with the source of the information or, technically, become the source.

8

Labels and Instructions: The Duty to Warn

In the effort to develop a successful product liability prevention program and ensure product safety, one must understand the different types of liabilities that the manufacturer becomes exposed to. Product liability claims are generally listed under one of two categories: *negligence* or *strict liability*.

Under manufacturing negligence, the courts will be deciding whether the manufacturer exercised due care in the ways in which the product was marketed or designed, or in the methods of manufacturing, inspection, or packaging. To prove that the design of the product was defective, counsel for the plaintiff will require a thorough understanding of the product itself, familiarity with state-of-the-art concepts and engineering practices, and a level of knowledge that begins to match that of the engineer who designed it.

Other than defective design, *lack of adequate warning* will be the next largest cause of product liability actions.To claim lack of adequate warning requires little or no product or engineering knowledge. In the legal field, it is almost a reflex action. The guidelines for when a manufacturer has the duty to warn comprise what are known as the elements of strict liability.

The Duty to Warn: The Elements of Strict Liability

Risk of harm that is inherent in the product or that may arise from an intended or reasonably anticipated use of the product.

Reasonable foreseeable or actually foreseen risk of harm at the time the product is marketed.

Failure to warn, failure to adequately warn, failure to provide instructions, or failure to provide adequate instructions regarding the danger.

Deficiency of warning or instructions rendering the product unreasonably dangerous.

Deficiency of warning or instructions that could cause user injury.

The American Law of Products Liability recognizes the following as the elements of negligence:

The Duty to Warn: The Elements of Negligence

Evidence of care regarding dangerous propensities of a product owed by the reasonably prudent person in similar circumstances.

Breach of standard of care (by failure to warn, failure to provide instructions, or failure to provide adequate warnings or instructions).

Injury caused directly or proximately by the breach.

Loss suffered because of the injury.

The first opportunity that an organization has to address these areas of concern is naturally when the engineer or designer first creates the product and determines the specifications. The first opportunity that the organization as a whole has to review these concerns is in the Design Review. In addition to evaluating the manufacturability and reliability of the proposed design, the Product Liabilities Review Board or Product Reliability Review Team also has to address the inherent risks that the product presents. Corporations that design and create few products probably put more effort into this and the development of the necessary instructions than do those companies that create numerous products or are considered job shops.

LABELS AND INSTRUCTIONS

Those in engineering disciplines and management become so familiar with their own product lines that they assume the inherent dangers and the elements of caution are common knowledge among the users as well. So, the more product that is created, the more the warnings and instructions become standardized, and actually begin to diminish. And one of the ironic aspects to much of this is that many products that were once used by members of certain industries or specialists are now being used by everyone. Look at computers as an instant example: where were they 10 years ago, and where are they now?

This level of expertise also becomes a problem when it comes to writing product instructions. Most companies would determine that the most logical individual to write an instruction sheet or manual for a product would be one of the engineers or technicians with the most technical knowledge of how the product works. Wrong. Who is likely to be the user of the product? Would it be someone with much knowledge and expertise, or would it logically be someone with little or no technical knowledge or expertise in the field?

Companies need to remember that most newspapers are written on a sixth-grade level—not because the writers went only as far as the sixth grade, but because that is the reading level of their average reader. Surely most reporters, if they wanted to, could write so that only college graduates could comprehend; after all, that is the level of knowledge most of them have. But then only about 15% of their readers would understand everything they have written, so what would they have accomplished?

Look back at the computer industry as another prime example of this problem. Software programs often come with manuals that are as thick as *War and Peace*. And when the average novice user tries to figure out how to use the program, he often quickly becomes lost. So he asks someone else who has the same program how certain things are done. Or, if another person isn't available, he either crawls through the operation and learns through trial and error or takes the easy way out and finds the technical-assistance number in the manual and calls the software people.

Engineers and technical specialists are not always the right people to write instruction manuals. Someone who is less familiar with the product or is less apt to write in a technical manner could be a better choice: someone who can translate technical into nontechnical and who has the ability to be a good writer but isn't literate in the technical area of the product. At the very least, once the instructions have been written, bring in the switchboard operator or someone from Personnel and see how successfully he or she can follow them without assistance.

PROBLEMS WITH PRODUCT EVOLUTION

But the problem with product development and design evolution is the same in many industries. Engineers and manufacturers keep putting out better versions and different variations of the same types of product lines, from TVs and VCRs to outboard motors. They begin to take a lot of the basics for granted, with respect not only to the development of the basic instructions but also to the basic warnings. Would a person who has just purchased his first outboard motor automatically know not to start the motor and run it while the boat is up on the trailer? It might seem to be a pretty stupid thing to a person who is knowledgeable about outboards, but would anyone else know that the motor would overheat? Probably not. And that is why manufacturers need not only to make sure that they don't drop their guard with instructions and warnings but even to improve their efforts.

Furthermore, warnings and instructions are not necessarily the same thing: "The former call attention to danger; the latter prescribe procedures for efficient use of the product and for avoiding danger. A manufacturer might provide one and still be liable for failing to provide the other, as where instructions fail to alert the user to the danger they seek to avert, or where a warning alerts the user to peril but does not enable him to avoid it" (Boyl v. California Chemical Co, 221 F. Supp. 669, 676 n. 6). Therefore, the first issue the manufacturer needs to address is whether instructions are needed; the second is whether there are warnings that need to be provided.

LEGAL GUIDELINES

With the Uniform Safety Act of 1987, Congress as a standard for determining a manufacturer's liability as it pertained to potential negligence endorsed the phrase what a "reasonably prudent person" might do. This is also the basis on which a court would decide when an operator or user was negligent for performing a certain action, as opposed to the manufacturer for not warning against it. A manufacturer normally won't be held liable for failing to warn about a danger that would be considered open and obvious, such as warning the owner of a powermower about the dangers associated with sticking a hand in by the blade (Ragsdale v. K-Mart Corp 468 N.E. 2d 524 Ind. Ct. App. 1984).

> If the manufacturer has greater knowledge than the average consumer would likely possess regarding pertinent safety information for safe use of the product, the manufacturer has a duty to pass that knowledge on to the consumer through a warning. Generally, a duty to warn exists where there is unequal knowledge, actual or constructive, and defendant, possessed of such knowledge, knows or should know that harm might or could occur if no warning is given [Miller v. Dvornik, 149 Ill. App. 3d 883, 501 N.E. 2d 160 1986].

Many courts subscribe to the presumption that, had an adequate warning been given, the user would have read and heeded such a warning. This is referred to as the *read and heed presumption*. If the plaintiff is successful in this attack on the defendant with the court and jury, the plaintiff is able to bypass causation (that is, having to prove any other defective condition in a product) by identifying that the actual defect was the lack of adequate warning. The origin of this presumption may be found in comment j of section 402A of the Second Restatement of Torts, which states:

> Where warning is given, the seller may reasonably assume that it will be read and heeded; and a product bearing such a warning, which is safe for use if it is followed, is not in a defective condition, nor is it unreasonably dangerous.

Therefore, if a defendant can take refuge in supplying what is viewed as adequate warning, then the plaintiff should be entitled to an opposite presumption.

CONSIDERING THE END USER

It is easy for a manufacturer to lose touch with reality and overlook the level of knowledge the average user of their products is likely to possess. It becomes almost a nuisance for the manufacturer and engineers to have to focus on this subject and the random possibilities. Yet, when the company enters into serious product liability litigation involving personal injury and is accused of inadequate warning, a major emphasis in this area is initiated to prevent potential future incidents. However, with the appointment of the Product Liability Expert and the creation of the Product Liabilities Review Board, it will become more of a routine for this issue to be addressed.

Determining which elements of caution and potential danger to address requires a lot of thought and consideration. Middle ground has to be sought between the electrical generator with no warnings and the stepladder with 110 warning labels. This can be an evolutionary process as the company gains more knowledge and experience in the field.

For instance, the manufacturer of a portable, metal-framed electrical display with more potential for producing electrical shocks than an average product started out with no warning labels or instructions. As years passed, the manufacturer became aware of infrequent minor accidents in which the users received severe shocks from the product—primarily the result of negligence on the part of the user but in-depth investigation was required before this was revealed. In the first phase of warning, the manufacturer placed a label on the product that read:

DANGER
HIGH POTENTIAL FOR ELECTRICAL SHOCKS

Misuse or abusive handling could result in severe shock. Maintenance or repair should only be performed by an experienced serviceman. Unit must be kept out of the reach of children.

As the years went by, fewer incidents were reported, and when a situation arose that was the result of a user's attempting to make adjustments of some sort to the product, the manufacturer was in the proper position to point out the warning that advised against such activity. But in a few cases, the user wasn't attempting a repair; somehow the product had been subjected to a physical motion or disturbance that altered its assembly and thereby created an unsafe electrical condition. When the user simply made contact with the product to adjust it while it was lit up on display, he received a severe shock. In the investigation of the incident, the Product Liability Expert could not explain how the defective condition had been created, and was unable to allege user negligence in simply handling and adjusting the powered-up display.

In trying to prevent future episodes, the manufacturer decided to expand on the wording on the warning label. It was changed to read:

DANGER
HIGH POTENTIAL FOR ELECTRICAL SHOCKS

Misuse or abusive handling could result in severe shock. Maintenance or repair should only be performed by an experienced serviceman. Do not make physical contact with any part of this display, without first disconnecting it from its power source. Unit must be kept out of the range of children.

Years later, another, similar surfaced in which an end user supposedly received a severe shock from merely adjusting the display as it hung on the wall. The Product Liability Expert met with the user's legal counsel and pointed out the red warning label on the product with the new wording advising against any physical contact. The attorney never pursued the case any further. As years passed, and the industry developed newer and safer electronic concepts replacing the earlier electrical designs, the manufacturer was able to change the product to such a degree that it no longer posed a potential for such danger.

In that example, the function that the product performed basically outweighed the potential risk, and with the proper warnings the manufacturer stood a good chance of defending the product. The attorney for the plaintiff could have still tried to pursue the case, alleging that the product was *unreasonably dangerous for the marketplace*, but it obviously presented too much of a challenge and expense to pursue.

In attempting to balance a product's utility against its potential risk, a company may find it helpful to consider the seven factors developed by Professor John Wade (On the Nature of Strict Tort Liability for Products, 44 Miss. L.J. 825).

1. The usefulness and desirability of the product—its utility to the user and the public as a whole.
2. The safety aspects of the product—the likelihood that it will cause injury, and the probable seriousness of the injury.
3. The availability of a substitute product which would meet the same need and not be as unsafe.
4. The manufacturer's ability to eliminate the unsafe character of the product without impairing its usefulness or making it too expensive to maintain its utility.
5. The user's ability to avoid danger by the exercise of care in the use of the product.
6. The user's anticipated awareness of the dangers inherent in the product and their avoidability, because of general public knowledge of the obvious condition of the product, or of the existence of suitable warnings or instructions.
7. The feasibility, on the part of the manufacturer, of spreading the loss by setting the price of the product or carrying liability insurance.

In a hypothetical situation, a manufacturing company created a set of decorative electrical lights that initially were designed for indoor or outdoor use. But through testing it was found that the lights posed a serious threat if used outside while it was raining, so the manufacturer decided to place a label on the product warning against allowing the product to get wet. If the product was marketed as an indoor or outdoor product and clearly presented as such, the courts may rule against the manufacturer regardless of the stated warning, in recognition of how the product was being marketed for use.

A similar argument could be made of the cellular car phone. The purpose of the car phone is to be enable the user to place and receive calls while on the road, thereby avoiding the need to stop to use a pay phone. The potential dangers of a driver's talking on a car phone while he should be concentrating on driving led some car phone makers to stipulate in their instruction manuals that the phone was not intended for use while driving. To place or receive a call, the driver is expected to pull off the road and stop. Although the car phone makers may try to use this documented warning in their defense, it has yet to be seen how the courts will respond in a case where being on the car phone is cited as the cause of an accident.

When a product is marketed within a special trade or profession, and the potential dangers are considered common knowledge within that trade or profession, the manufacturer may have no duty to warn, even though the dangers may not be known to a person outside that group. But the manufacturer of a piece of equipment used in industry may still have a duty to warn through the use of instructions about the potential dangers inherent in the machine, even though the end users may be considered experienced.

DEVELOPING WARNING LABELS

When developing warnings or warning labels, one must determine the best way to present the information so that it will be noticed and understood by the end user. Guidelines for developing warning labels are contained in the ANSI (American National Standards Institute) standard Z535.4. The standard states that a product safety warning label should alert the person to a specific hazard and indicate the degree of risk, the consequences of involvement when subjected to the hazard, and how to avoid the hazard (see Figure 1).

With regard to the placement of the warning label, the standard states that it should be distinctive on the product; be located in the immediate vicinity of the hazard, readily visible so the viewer can recognize the hazard and take appropriate action; and be designed with the expected life of the product and the foreseeable environment in mind. In developing the label, the designer should use one of the following signal words:

DANGER: Denotes an imminently hazardous situation that, if not avoided, will result in death or serious injury. The signal word *Danger* should be printed in white lettering on a red background. (All colors specified for use on the product safety signs and labels should conform to ANSI Z535.1-1991.)

WARNING: Denotes a potentially hazardous condition that, if not avoided, could result in death or serious injury. The signal word *Warning* should be printed in black lettering on an orange background.

CAUTION: Denotes a potentially hazardous situation that, if not avoided, may result in minor or moderate injury. The signal word *Caution* should be printed in black lettering on a yellow background.

In addition to the signal alert words, the warning should include a safety alert symbol, a symbol/pictorial panel, and a word message.

The safety alert symbol: A triangle that immediately precedes the signal word. The solid triangle portion must be the

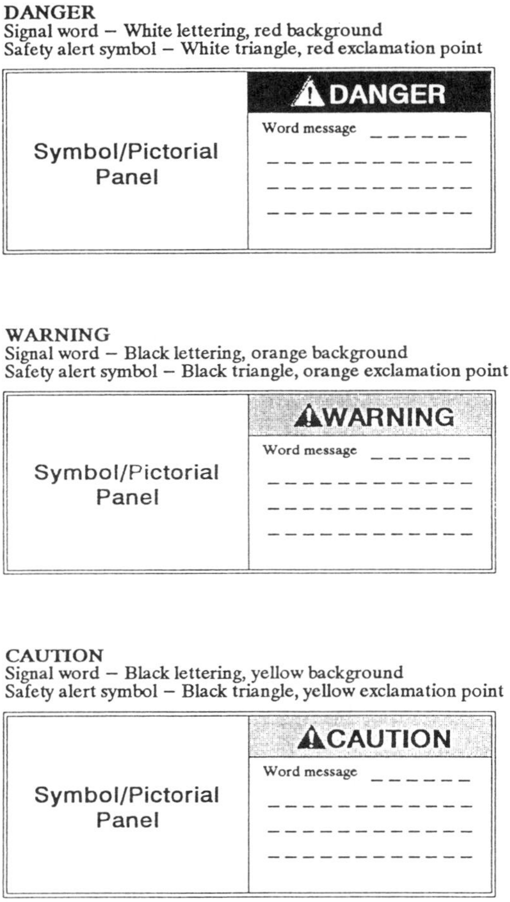

DANGER
Signal word – White lettering, red background
Safety alert symbol – White triangle, red exclamation point

Symbol/Pictorial Panel
DANGER
Word message ______

WARNING
Signal word – Black lettering, orange background
Safety alert symbol – Black triangle, orange exclamation point

Symbol/Pictorial Panel
WARNING
Word message ______

CAUTION
Signal word – Black lettering, yellow background
Safety alert symbol – Black triangle, yellow exclamation point

Symbol/Pictorial Panel
CAUTION
Word message ______

Figure 1 Layout of hazard label. Pictorials: black symbols on white background; word messages: black lettering on white background or white lettering on black background.

same color as the lettering used for the signal word, and the exclamation mark contained within the triangle must be the same color as the panel background for the signal word. For example, the safety alert signal that accompanies a *Danger* warning should be a white triangle with a red exclamation mark.

The symbol/pictorial panel: The pictorial panel should be easily understood and should clearly and effectively convey the message. The pictorial panel generally should have a black background. Additional colors, such as red, may be used for emphasis.

The word message: The written message should be concise and easily understood. If detailed instructions, precautions, or consequences necessitate more extensive discussion, it is permissible to refer the user to the instruction manual. The word message should have either black lettering on a white background or white lettering on a black background.

It should be recognized that the ANSI standards have been established as a recommended guideline for designing labels but do not have the force of law. In that sense they are much like the UL standard for construction, which is the recommended standard for state-of-the-art safety but there is no requirement for conformance. However, it is in the manufacturer's best interests to demonstrate compliance to the standards, just as with the UL guidelines, or it will become a plaintiff's advantage to make a point of the noncompliance.

THE CONTINUING DUTY TO WARN

In addition to the initial duty to warn, if the manufacturer becomes aware of an unsafe condition existing in the product even after the sale, they may bear a continuing duty to warn. It is recognized that at the time of manufacture or sale the manufacturer can be held liable only for information that is known at that time. But as the manufacturer gains more knowledge of the product and its potential hazards over time, some courts have imposed that the manufacturer has a responsibility to share or communicate that knowledge with the purchasers or users of the product. Such knowledge may be gained through additional testing or industry studies that lead to an improvement in the state of the art or through accident reports received by the manufacturer.

Just as with the initial duty to warn, the continuing duty to warn is based on what would be considered "reasonable." The nature and severity of the hazard, the burden that it would impose on the manufacturer to locate the end users, the attention that one could expect from any type of reasonable notice, the nature of the product and the quantity that was produced and circulated are all factors in the determination of what would appear reasonable and practical. This becomes even more complicated when the original product was sold to various distributors and so the records don't show where the products went from there, or in situations with equipment and machinery that may have been resold from company to company.

In some states (e.g., Massachusetts), the courts have declined rulings regarding the manufacturer's continuing duty to warn, and in others the duration required remains unclear.

> It is beyond reason and good judgement to hold a manufacturer responsible for a duty of annually warning of safety hazards on household items . . . when the product is 6 to 35 years old and outdated by some 20 newer models equipped with every imaginable safety innovation in the state of the art. It would place an unreasonable duty upon these manufacturers if they were required to trace the ownership of each unit sold and warn annually of new safety improvements over a 35 year period [Kozlowski v. John E. Smith's Sons Co., 87 Wis. 882, 275 N.W. 2d 915, 923–34, 1979].

If a manufacturer does recognize the duty to warn based on newly acquired information, they can carry out this responsibility in several ways. The first and foremost is to contact end users directly and supply them with the necessary warnings, instructions, or safeguards. A second way is through direct mass mailings that communicate the newly learned hazards and give specific instructions. A third way could be through the use of notices in specific industry or direct-focus publications. It needs to be mentioned that this is all in addition to the responsibilities that a manufacturer may have regarding the laws and regulations imposed by the Consumer Product Safety Commission, as explained in more detail in Chapter 11.

FORESEEABLE MISUSE

As with reliability testing, the next level of consideration in providing warnings is the area of foreseeable misuse. The product labels, along with the accompanying instruction sheets, now address all the known warnings and potential hazards as foreseen by the manufacturer during normal operation, but now we want to explore the possibilities of foreseeable misuse and determine if there are other practices that should be warned against. The objective nature of this to some extent makes it difficult to explore, but there are common abuses almost any product might be found to be susceptible to—a very general example would be hitting a product to try to make it work, and yet technically it is foreseeable misuse or abuse.

It is very common for people to cut the grounding prong off extension cords and other products with three-prong plugs, which significantly increases the potential danger of the product. Vending machines become like slot machines when it comes time to see if a selected product will really drop down. And if it doesn't, how many upset customers will begin banging on the machine to make the product drop, or begin rocking it, which could easily result in the machine's falling over and injuring someone. In this day and age, new cars aren't really intended to be used for jump-starting the batteries of other new cars, and yet it is entirely probable that they will be; therefore there should be a warning against it under the hood or the owner's manual should describe how to do it properly.

There is naturally a point at which abuse becomes negligence, but the two points may be far apart from the perspective of the design engineer as opposed to counsel for the plaintiff. During the Design Review, the Liability Expert and the rest of the organization needs to reach a logical middle ground where the foreseeable is identified and addressed, without entering into the ridiculous.

RIDICULOUS BORDERING ON HUMOROUS

One excellent example of a possibility that most people would overlook concerns a simplistic product, the Clapper (Hubbs. v. Joseph Enterprises 604 N.Y.S.2d 292, 3rd Dept. 1993). An 80-year-old arthritic woman sued the manufacturer of the renowned device, claiming that she was injured due to "extra hard clapping." The plaintiff contended that her injuries were the result of the manufacturer's failure to warn against the hazard. But the manufacturer actually *did* warn the user in the literature accompanying the Clapper that it might be advisable for the user to take advantage of a clicker as an alternative to actually clapping. The court found that the plaintiff's claims were meritless, since she obviously chose to ignore the literature, and in fact made one last deliberate attempt to activate the unit with "one last hard clap" when other repeated efforts had failed. The court dismissed the complaint.

9

Records Retention

When a manufacturer enters into the litigation stage of a product liability action, one of the elements that becomes paramount in working with the other parties and the courts, and primarily in putting together the defense, is records retention—being able to factually describe and display exactly what was happening at the time the product in question was being designed, engineered, marketed, sold, tested, manufactured, and shipped. As the case develops, factors surrounding these events will be thoroughly analyzed and critiqued.

Counsel for the plaintiff will be looking for evidence to substantiate that the company in question may have demonstrated negligence, recklessness, or carelessness in the design, manufacture, or distribution of the product prior to releasing it to the marketplace, in short, that the company demonstrated insufficient effort or concern.

The manufacturer, on the other hand, will be trying to prove that they are, and were, a caring company at the time this product was developed. They will try to prove that they made every reasonable effort to ensure a safe and reliable product, and complied with all aspects of relative specifications, codes, and standards. This will require a lot of documented evidence.

The key statement here is "at the time the product was developed." Most products that are involved in some type of product liability incident, whether fire, accident, or personal injury, are not likely to be fresh off the assembly line. In many cases, the product will have been produced 3–5 years prior to the incident. And there is where the Product Liability Expert begins to experience problems. Companies are normally good at having immediately at hand various documents that relate to the here and now, but not the past and gone. It often seems that once a product or product order has been produced, shipped, and billed, it becomes ancient history. The focus then shifts to the present, as no one seems to hold much concern for the past, especially what might be viewed as the distant past. So the records depicting past actions may be packed away in unknown locations, if they continue to exist at all. But product liability tends to deal more with the past, and the manufacturer had better know—and be able to prove—exactly what happened.

SALES

One of the first areas of attention and concern will logically be the sales or purchase order. When was the product purchased? When was it shipped or delivered? When was it installed or put into action? What was specified? What was the market application? What were the contractual agreements? Who had what responsibilities related to sales, service, distribution, installation, and performance? What were the contractual agreements regarding warranties and liabilities? These are just a sampling of many questions that will likely be asked and analyzed, and the defendant manufacturer has to be prepared not only to answer but to back up their answer with documented proof.

Sales staff and Account Managers tend to create files when they begin working with customers on specific orders. The files contain every bit of documentation and correspondence relative to the product and order, and can develop into a number of files as the order progresses. Eventually, once the job is complete, or the account person recognizes the need to purge files, the file with all this history is at best packed away somewhere to free up space, and life goes on.

In recognizing how important these files and documents are in the event of a future product liability incident, the company now needs to develop a system and procedure for the identification and retention of such records and files. The procedure should clearly identify where and how the records will be kept, as well as the retention period. If the company enters into a liability action, and counsel for the plaintiff submits a request for documents asking specifically for the original customer's purchase order for the product that ran 4 years ago, by following the instructions of this procedure the Product Liability Expert should be able to go directly to the specified records retention area and pull the records needed without the aid of Sales and Customer Service personnel. This should be achievable without having to dig through piles of dusty packages in basement or attic locations, searching numerous shelves and opening countless vaguely described box files, or taking any other laborious measure to get the needed material. Many companies would have to make a real concerted effort to get their records retention practice to this level. But this is where it needs to be.

ENGINEERING

Just like Sales, Engineering tends to be more concerned and involved with their current projects than with past projects. Products that were designed, engineered, and manufactured 3–5 years ago are likely filed away. The current job being engineered is on the table (or tube), and the most recent predecessor is folded up somewhere on the desk because manufacturing may still be having problems with it. But the jobs that were engineered years ago have already been manufactured, the drawings subjected to numerous possible revisions, and are now filed away.

The problems experienced here can be a little different from those described with Sales. In all probability, engineering drawings are easier to find and are probably right there in the department, rather than having been moved elsewhere. This makes it a lot easier for the Product Liability Expert trying to get copies of all the drawings associated with a specific product. The problem that is unique to Engineering is that the product on file may be at revision K or P while the product in question was manufactured in compliance with revision A or B. Although in most systems engineering changes are clearly documented, allowing a reviewer to read the evolutionary history of a product right up to the current time, there is still a problem. If the product has undergone massive changes since its conception, the reviewer may not be able to lay his or her hands on a drawing of the actual product at the time it was manufactured 3–5 years prior. The reviewer, or the Liability Expert, would have to take a current print and eliminate all the changes that were incorporated into the product after that date, in order to conceptualize the engineering of the product for a specific period.

As part of this effort, Engineering must now keep a copy of each drawing at every stage of significant revision. If the drawing was now on revision C, the Expert would have easy access to the drawings done under revisions A and B. If counsel for a plaintiff wanted a copy of the revision A drawing, the Expert could provide it without ever showing revisions B and C, not to mention everything up to M or T. Note that copies need not be kept reflecting literally every modification. If the changes introduced are minor (e.g., screw heads changed from being slotted to Phillips-style or the color changed from black to blue), they can be appreciated without maintaining separate prints of each version. But when the configuration changes or the product goes through an evolutionary upgrade that may make it more difficult to imagine exactly what the product used to look like, then we need to be able to have the previous drawing at hand.

This becomes even more crucial when dealing with labels and instructions. The primary product drawing may reference labels or associated product instructions only by the number assigned to the form itself. A copy of the actual form is found under its drawing number. If it is a one-page instruction sheet, or a one-page drawing of the label, the files may contain only the current version, since the past versions are no longer viewed as being relative. If the Expert or the plaintiff's attorney wants to see exactly what the labels or instructions read at the time of the incident, and they definitely will, there may no longer be a copy of them. That would lead to real problems.

Finally, if a company allows an engineer to make changes to drawings and specifications without complete documentation, this is a major problem. In some operations, engineers may view documenting changes as just a time-consuming formality. They feel that if they agree with the need for the change, then that is all that is necessary. This kind of practice has to be stopped immediately. If products go through evolutionary changes and upgrades without any documentation or recorded revisions, the Liability Expert will have no idea how the product was actually manufactured at a specific point in time, and that will have a devastating effect on the manufacturer in court.

There is another concern to think about regarding drawings and revisions. When a product liability case surfaces, counsel for the plaintiff is always going to request a drawing of how the product in question was supposedly designed and manufactured. If the only copy maintained reflects the current revision M, you may not want the other party to see all the changes that transpired since the product in question was built under B. If the changes improved the safety and reliability of the product, it will demonstrate to the courts that even the manufacturer recognized deficiencies in the design and corrected them. This isn't to say that the plaintiff won't eventually ask for subsequent drawings anyway, which defense may object to, but the information doesn't need to be made known immediately.

Another real problem that we face in this computerized age is that more and more engineers are no longer at a drawing table but in front of a tube. Computerized drawings make the engineer's job easier and more accurate. They are also easier to store, because they don't have to be filed as paper: at the time the product is manufactured, or anytime someone is in need of a current drawing, the engineer can generate a copy right off the computer. Once the drawing is used, it is usually discarded. This ensures that the user is always working from the latest version rather than pulling from drawers what may be obsolete prints. Once again, then, there is the problem of not having at your disposal a drawing of the product at any specific stage. The company has a few options to deal with this dilemma. First, you could print and file a hard version of each drawing prior to any significant revision. Second, you could retain the previous versions on the hard drive; for instance, the file would read part nos. 3791A, 3791B, 3791C, etc. In this manner you could print out any of the old versions.

The bottom line is that you as the manufacturer have to be in a position to show exactly how the product was built at a specified time, without offering anything more. Your system must therefore be designed accordingly.

OTHER QUALITY RECORDS

Although engineering specifications, sales and purchase orders, and contractual agreements are the most frequently requested items in any product liability suit, numerous other documents may also have to be shown as the case progresses. It is wise for any manufacturer to have a records retention

Dept Resp	Record Description	Retention Period: 6 Months	1 Year	2 Years	3 Years	5 Years
QA	Obsolete Quality Procedures					X
QA	Obsolete Process Instructions					X
QA	Receiving Inspection Records			X		
Purchasing	Supplier Rejected Material Reports			X		
QA	Audit Inspection Records				X	
QA	Customer Product Complaint Forms				X	
Engineering	Previous Blueprint Editions					X
Engineering	Engineering Change Notices					X
QA	Production Audit & Product Testing Reports					X
QA	Product Reliability Testing Reports					X
Purchasing	Purchase Orders to Suppliers					X
Cust Service	P. O.'s and Spec's from Customers (after completion)					X
Cust Service	Purchase Orders to Subcontract Companies					X
Purchasing	Purchase Orders and Specifications to Subcontractors				X	
QA	New Product Introduction Minutes				X	
QA	Supplier Performance Evaluations			X		
QA	Procedure Audits					X

Figure 1 Quality records retention chart.

program in place that spells out what types of records need to be retained, the party responsible for retaining them, and the retention period.

In general, there are various guidelines for record retention that have been established for different industries. For instance, the U.S. Consumer Product Safety Commission has established guidelines for companies involved in the manufacture of consumer products which stipulate that most records need to be maintained for at least 3 years. Likewise, the FDA has prescribed periods for those involved in the manufacture of medical devices.

As part of the Quality system, a chart should be created that shows the types of records to be maintained, along with the duration period. An example is shown in Figure 1. This simple chart clearly identifies all the various records deemed important, along with the department responsible for keeping them, and the retention period. These retention periods can vary from company to company, as can the length of the list. In some industries, a company may find that they need to retain certain records for 7–10 or even 20 years after the product is produced.

As part of this new focus, the Product Liability Expert and the Review Board need to recognize all the types of documents and records that could be critical to their defense, and establish retention periods for them.

10

Dangerous Documents: The Smoking Gun

In launching a corporate effort to prevent the possibility of entering into potential product liability, as well as to help successfully defend the company against actions that do arise, a few major practices need to go into effect almost immediately. The Quality system and effective Design Reviews are two of the most primary initiatives, along with learning how to—or not to—write documents and memos. Unfortunately, most corporations initially learn about this third concern through firsthand experience, when they're actively involved in a lawsuit. Then, of course, it is too late.

Certain documents lying around in the corporation's files or various types of letters that have been sent out can end up being the nails that are driven into the corporate coffin when discovery is initiated. It is commonplace for every manager or company executive to think that his or her files are confidential, and that no one will ever see the contents without the individual's permission, but this is all untrue. When a product liability incident surfaces and counsel for the plaintiff demands copies of certain documents, memos, or other forms of communication, it isn't a mere request; it is a court order to present them. Failure by any individual within the company to comply would be viewed as contempt of court. It also needs to be understood that such documents that were created up to the point of litigation are not protected under attorney–client privilege.

DEFINITION

Dangerous documents are those that identify that at some point the company gained knowledge or became of the opinion that the product was unsafe or unreliable, and other records may show that little or nothing was done about it. Or they can be documents that discuss a better way to design and manufacture the product, when in fact the final decision was to pursue a less expensive and conceivably less safe or reliable concept. This is how a document may become a "smoking gun." (While searching the home of a murder suspect, the police find the gun used in the killing lying on the counter, the barrel still smoking.) Finding documents of this sort in the ofices of the manufacturer proves that the company had prior knowledge of a certain condition. Basically, a dangerous document is anything that the company wouldn't ever want to have presented in court, or have to stand behind and defend if it became part of the product liability evidence.

Companies tend to create dangerous documents as part of several different functions or efforts: in the initial sale or quotation process and discussion, in the design or engineering phase and especially in the Design Review, in the production or Quality Control phase and effort, or in the many reliability tests or ongoing communication and correspondence with the customers or end users. Such documents can be created at almost any time during the product's life cycle, and then lie waiting to be discovered as part of the historical record.

Sometimes they are created as part of a legitimate concern regarding the proposed or current design of the product, for example, by an executive staff member voicing concerns over a product, potential warranty costs, or customer approval. Or a document may go as far as discussing possible liabilities and hazards. But if the company finally opts not to go with the individual's recommendations, the related documents have the potential to be dangerous.

Sometimes they are created as part of the corporation's internal ego system, other times as part of an individual's cover-your-backside (CYA) effort. In many circumstances they are derived from speculation and conjecture, as opposed to being based on factual information or results. For instance, it would not be unusal for a junior member of management to initiate a memo voicing an opinion on some element of a product's design or manufacture just to gain attention or to have been the first to have advised against the concept prior to its potential failure in the field.

E-MAIL: THE DOCUMENT THAT NEVER DISAPPEARS

As most companies move into the computer age with terminals at every workstation, they enter into an even more complex situation in relation to dangerous documents. With ease, employees can now fire off electronic memos to a distribution list of others within the organization, without an actual hard copy existing. Yet the correspondence does legally exist, and can be demanded.

This situation becomes further complicated because, even if the sender and receiver deletes their files, the system's mainframe typically downloads its memory each night, which means that the document may still exist, and could conceivably reappear. Computers now add a whole new dimension to the terms *documents* and *files*.

PREVENTION THROUGH TRAINING

The Product Liability Expert needs to teach all the employees within the corporation what types of document should never be created in the first place. These are some examples:

Documents that try to place the blame elsewhere for a problem related to the product, as when an employee or department recognizes that they are being blamed for a problem that exists with a product and they try to shift the blame to another department or individual.

Documents that discuss the cost of improving the safety or reliability of the product, stipulating how much more per unit it will cost to improve the design and how it will affect profitability or competitiveness.

Documents that discuss speculative or potential product liability or product reliability concerns, as opposed to factual results or findings.

If the company reacts to an opinion or recommendation and makes the required changes to the product, then the memo or other form of communication may not become a dangerous document: someone recommended that an improvement be made for whatever reason and other documents show that the change was actually implemented. But if the company doesn't respond, or can't show that a positive reaction transpired when the concern surfaced, then the document could help substantiate negligence.

ADDRESSING DESIGN PROBLEMS

In many product liability cases that have achieved national attention, opposing counsel presented as evidence engineers' test reports that demonstrated that management had prior knowledge of the condition and chose not to react. Sometimes the individuals that make these documents known are no longer with the company but have taken their personal files with them. In other cases, internal memos that were circulated among management discussing product concerns and ultimate decisions became a major point of embarrassment to management. Whether the case deals with the automotive industry or the space industry, or even as high as the White House, documents and memos are always among the key factors that do the defendants in.

Corporations will always have issues and concerns that need to be brought to the table and discussed, but what they need to learn is how to properly address them. Take a hypothetical situation in which a company wants to introduce a new VCR to the marketplace. In the initial testing of a prototype unit, it is found that if the unit experiences an electrical surge due to lightning a certain programming chip located on the circuitboard has the potential to pop and may actually sustain a small flame that has the slight possibility of spreading and eventually leading to a fire. This condition is known only by the Reliability Testing Engineer, who feels that it is vital information to bring to the management group's attention.

The engineer knows that the immediate alternatives for the manufacturer are: to incorporate a surge suppressor in the circuitry that will cost an additional $5 per unit; insert a fuse in the unit that will blow when the surge happens, for an additional $2.75; replace the programming chip with another type that costs $4.25 more per unit; or recommend in warnings and instructions that the unit be plugged into a surge suppressor outlet strip or box, at the user's expense.

In bringing this concern to the table, the Reliability Engineer takes advantage of an already scheduled Design Review meeting or Product Introduction meeting, or decides to call a special meeting with the principal parties involved in the launch of the new product. Up to this point, nothing has been documented regarding these findings.

MEETING MINUTES

At the meeting, the Reliability Engineer communicates the findings to the group, along with the alternatives. The group determines the risk factor involved here, along with the various costs. Let's say that the company decides to incorporate the surge suppressor into the circuitry for an additional $5. The meeting minutes then reference the test findings, along with the improvement being pursued. Even if the chances of shock were only one in 10,000, the potential for failure has been addressed and eliminated.

Let's say that in another hypothetical course of action, the same findings are brought to the table and the group once again weighs the risk factors. This time, because making the changes seems too expensive and the perceived risk is felt to be so remote that the company decides to place the applicable warnings and instructions on the product and in its manual. The meeting minutes then briefly mention this hazardous condition along with what the company is going to do to properly warn the user and hopefully prevent the hazard. But nothing is documented regarding the other alternatives or their costs.

At a later date, an incident involving the product happens in the field, litigation begins, and counsel for the plaintiff pursues discovery. The Design Review records note that: "A concern was expressed regarding the remote potential for fire if the unit experienced an electrical surge. Although the chances were considered extremely remote, the group decided to act and that it would be brought to the attention of the user as a warning that would be applied to the unit, as well as be incorporated into the instruction booklet." This exhibit would now demonstrate to the court that the company considered this condition and, even though they found the possibilities to be extremely remote, they decided to act on it and incorporated the various warnings and instructions.

Since the test results weren't documented and aren't being mentioned, nor are the improvements or cost considerations, the company in this last scenario would still be viewed as acting responsibly. Had all this other information been mentioned as in the previous scenario, the courts could have found that the company was negligent in its actions, and didn't do enough to prevent the possibilities. And it would have been the minutes from this meeting that would have been the prime piece of evidence.

Let's take the same hypothetical situation, and again have the Reliability Engineer present his undocumented findings to the management group in a special meeting. The group once again considers the risk along with the alternatives, but in this scenario decides to do nothing because of what is viewed as the remoteness of the possibility. And in this scenario no meeting minutes are recorded. Two years later, an incident does transpire and, as before, counsel for the plaintiff pursues discovery. This time there is no dangerous document to be found.

Even if a disgruntled ex-employee is brought into the picture and testifies that the problem was known and presented, without documented evidence or other witnesses to support this allegation such testimony could be viewed as vindictive and ruled out as unsubstantiated evidence. It becomes a serious problem only when memos and other documents enter the scene. This last meeting technically never took place. Furthermore, if witnesses are questioned during deposition as to whether this condition was ever discussed, the courts could not expect anyone to remember what may have been said at a brief meeting two years before (e.g., President Reagan disputes Colonel North's recollections of an alleged meeting to plan the Iran–Contra deal).

Although this example may be a little facetious, the point is that companies are going to have potential product liability problems brought to their attention that they will be forced to deal with, and the appropriate course of action won't always be easy to determine. The company will always be responsible for its own product, and will therefore have to make a responsible decision. The key is to learn how to document these decisions, and how not to. The ultimate responsibility of the company will always be there.

SUPPRESSING RECORDS

One serious aspect of not reacting to a known defective condition, especially if there are documents that do exist but the company decides not to offer them during discovery or any other investigation, is the possible repercussions the company could face in the hands of the government. If, for instance, during an investigation a whistleblower makes it known to government agencies or the courts that a company was in fact aware that they had a defective product in the field but appeared to be covering it up, the government could react severely by charging the company with suppressing records or information, and fraud. This could result in criminal indictments and fines in the hundreds of thousands of dollars, or even beyond a million dollars.

It is not the intention of this book to imply in any way that a company should ever consider a cover-up or a withholding of records. The company must put every effort into assuring the public safe and reliable products, and must react in the most effective manner when addressing these known concerns and variations. But the company must also quickly learn to put a stop to negative communications that might pose numerous problems later forcing them to consider the

DEPARTMENTAL CORRESPONDENCE

TO: Director of Engineering

FROM: Director of Quality

SUBJECT: Proposed Design Changes to the Model XB

In a routine reliability test on the Model XB where we subjected it to temperatures above 125 degrees F, we found that the internal power coil not only overheats, but during the process it begins to short-out and eventually has the potential of starting the unit on fire.

We have over 7000 of these units out in the field, and a large number of them are in the Southwest, where they have a real possibility of experiencing these temperatures! I don't know why these units weren't thoroughly tested by your own department prior to being produced, but I am going to insist that a thermo-protector switch be designed into the unit that shuts this unit down at 110F. An Engineering Change Request will be put through shortly.

I hope we don't hear about any fires in the field, or this will become a major problem.

cc: President
VP - Operations

Figure 1 Example of a dangerous document.

alternatives to either get rid of them (destroying evidence) or hide them (suppressing evidence), both of which would be criminal acts.

With group sessions, the Product Liability Expert can begin to teach the members of the organization the difference between effective documents and dangerous documents. Examples of documents that probably already exist can be pulled from circulation, copied, and presented as part of a training demonstration. As an example, Figure 1 shows that a defective condition was identified in a company's model XB through some routine testing. The Quality Director, obviously upset by what could be summarized as being an inadequately designed unit, makes it known that he will take it upon himself to straighten out the situation on behalf of the organization, and makes sure other important staff members are aware of it.

With this document in at least four files now, not only does the company need to be concerned about the engineering change to the product from this point on but they have also created a liability regarding the 7,000 units in the field. If a year from now one of those 7,000 units were to create a fire, and possibly a related injury as well, and through litigation and the discovery process counsel for the plaintiff were to gain access to this memo, the company and its management could be found liable and negligent for not making an attempt to warn the users.

This document also displays an apparent problem between the ranks, and possible lack of system compliance or initial design evaluation. This too could create negative impressions of the company during trial. It could eventually prove through cross-examination that various departments aren't following their own Quality system, that one department is being pitted against another, with the whole company standing to lose.

In Figure 2 we see the same message being communicated, but in a nonlethal manner. Here the Quality Director is describing an improvement that is being recommended, without getting into a lot of detail or referring to potential problems. It is also being recommended that the Director of Engineering call him and discuss the matter if he wants to, which effectively allows for the details to become known and the concerns shared in a verbal manner, which is just as effective, without creating the potential for having a dangerous document on file.

This memo is also short and to the point, which all memos should be. In Figure 1 the Quality Director goes beyond communicating the point at issue, and goes on to make other points known as well. An important rule of thumb is

DEPARTMENTAL CORRESPONDENCE

TO: Director of Engineering

FROM: Director of Quality

SUBJECT: Proposed Design Changes to the Model XB

I would like to propose a design improvement to the Model XB. I think we could improve on the units' performance and reliability if we designed into the unit a thermo-protector switch that would shut the unit down, if it were ever to experience an operating temperature over 110F.

I will be initiating an Engineering Change Request in order to incorporate this switch, which you will receive shortly. If you would like to discuss it further, please give me a call.

Figure 2 Memo addressing the same situation, worded in the careful manner of an effective document.

when you're in an emotional state . . . don't write memos. Wait until you have settled down, or even until the next day, before addressing an emotional issue on paper. In most cases it wouldn't make much difference whether you communicated certain issues or feelings today or waited until tomorrow. But most letters and memos written in a high state of emotion are regretted the next day, when it is too late.

DANGEROUS FORMS

Any type of standard form used as part of the internal Quality system could turn out to be a dangerous document, such as the Quality Hold Notice shown in Figure 3 or a Product Deviation Notice or Form. Because such forms are part of the system, they may be even more likely to be retained. The fact that a defective condition was found in the products being produced isn't a problem, nor is the fact

QUALITY HOLD NOTICE

PRODUCT: Electronic Converter **PART NO.** RK-27 **DATE:** 3/25/95

ORDER SIZE: 7200 . **QTY. IN STOCK:** 1500 **QTY. SHIPPED:** 1100

DEFECTIVE CONDITION: Sample units inspected in assembly were found to be miswired. The condition will probably lead to the unit shorting out, and possibly catching on fire, or becoming electrically live.

IMMEDIATE CORRECTIVE ACTION: All units assembled for the rest of the order will be audited to ensure that they are properly wired, and labeled with an orange circle.

CORRECTIVE ACTION TO INVENTORY: All packaged units in the warehouse have been tagged for rework, and will be brought into Area #2 to be rewired. All reworked units will have an orange circle applied to the as well, signifying that they have been reworked and audited.

FIELD CORRECTIVE ACTION: There isn't much that we can do about these units. When we receive calls from the customers about them, we will just have to arrange to have the defective ones returned and ship them a replacement.

DISTRIBUTION: Operations Manager
Director of Engineering
Assembly Superintendent
Assembly Supervisor

FROM: Quality Manager

Figure 3 Example of a standard form that could be a dangerous document.

that some of the stock had to be reworked. There isn't even a problem with the fact that the form describes the potential effects of the defect. The problem is that it states how serious the potential effects are, and the fact that nothing will be done regarding the products in the field.

Suppose one of the 1,100 RK-27s in the field does create an incident one to three years later, and through the request for documents counsel for the plaintiff comes across this form. This would clearly demonstrate negligence on the part of company management and incriminate the company. But if the document stated that the units would be recalled or handled in some other effective manner, it would demonstrate to the courts that when the defective condition was made known, the company reacted in a very positive manner. And this would be the case even if the product was still involved in some other product liability incident.

The key point isn't that you should hide your mistakes. Everyone is going to make mistakes, even with the best programs and efforts in place. And the courts recognize that. The key is to react to those mistakes immediately, and effectively, and be able to prove it. As long as a company can produce good records of how products were produced and checked, and how variations were handled, it is in good shape.

11

Warranties and Recalls

WARRANTIES

When a company begins to add up the costs of field warranties and possible recall, it is dealing with the costs of failure. These are the products that eluded the company's best efforts in the areas of prevention and appraisal and made it to the customer. Warranty costs may, in the best of situations, be minimal and in the worst case drain all the profits originally made on the product or job, and beyond. Recall costs are quite commonly staggering.

The first issue that a company needs to address is what its product warranty will be. In most situations, the general warranty coverage may be common knowledge, but it isn't documented anywhere. There may not even be a form that goes along with the product, which means that everyone receiving the product has to call to find out what's covered and what's not. It may be generally known that the products carry a one-year warranty against defects in workmanship or for mechanical function, but that is basically it. And the warranty that exists, if it is documented, may have been created years ago and have little relevance to today's products.

Customer Service: Interpretations

An element that is more of a day-to-day concern is Customer Service's knowledge and understanding of the warranty. These are the people whom all these calls are going to come to, and yet it may not be clear even to them. Many times the documented warranty is so brief and so full of disclaimer clauses that it really doesn't say very much. When issues arise and the Customer Service representatives receive phone calls that describe technical aspects of the product and problems being experienced, the issues tend to fall into gray areas with respect to whether they are covered by the company's warranty. Customer Service personnel have to discuss the matters with other management to find out whether the company wants to stand behind certain defective conditions, and allow for return, replacement, or service. And, after a period of time, these same people almost always have to find out what the company's attitude will be toward certain problems customers are facing.

In some cases, although the company may not normally replace or rework a product, they will if the customer yells loud enough. In doing so, management demonstrates to the Customer Service representative that they will stand firm on their policies . . . well, unless it becomes a problem. In a short time the Customer Service representatives learn to just tell management right from the start that the customer is really upset, thereby saving themselves the trouble of going back and forth or having to deal with an irate customer. Management's strategy is to keep Customer Service from "giving away the farm," so they start out by saying no to everything, and later give in to most things even though they think they are keeping tight pursestrings.

Unanimous Agreement

The company needs to spend time addressing its warranty and getting everything as clearly documented as possible. This is another area where the Liability Expert can take a lead role. Usually this requires collecting everything that currently exists, which may be the current documented warranty form, contractual agreements with various customers, letters and memos dealing with the subject, and the records of the situations that the company has commonly faced up to this point.

Although the company may not initially want to come right out and accept certain situations, if they recognize that they quite typically end up doing so anyway then they should just accept it and incorporate it into the official document. For instance, a company's product is a piece of machinery that involves painted metal and, although they warranty the mechanical function of the piece of equipment for 1 year, there is no mention of the paint. But if a customer who has had the unit for 7 months calls and says that the paint is beginning to peel, the company knows that it has to stand behind the product so the warranty should henceforth clearly state that the paint job is covered. Likewise, if the warrantee already states that the paint job on a car is covered for 1 year but you know that realistically you would end up standing behind a defective condition brought to light 13–14 months from purchase, then you should consider extending your warranty to 18 months or 2 years.

These items should then be clearly documented, not only for the customer, but for the company's personnel as well. Doing this should help reduce the number of calls from customers who don't know what the warranty policy is, and it will help the company's own personnel in dealing with the calls, and the subject in general. Many customer complaints arise from misconceptions about various aspects of the warranty: the terms and conditions, the length of warranty, and other specifics. This isn't to say that this effort will totally eliminate the gray areas, because it won't. There will always be some gray areas and bending of the policy, but it needs to be by exception and not as a rule.

Competing in the Marketplace

When addressing and establishing the new warranty, in addition to collecting everything that currently exists, the company should also collect that which is offered by the competition. In the competitive marketplace, the company that leads the pack isn't the one that makes a good product and just backs it for only 1 year; it will be the one that makes a good product and backs it for 2 years. It is also quite common that the company that charges less for the product but backs it for only 1 year will be less successful than another company that may charge a little more but stands behind their product for 2 or 3 years.

People want a quality product, and are willing to pay more for it. In a major survey conducted in 1991 for the American Society for Quality Control, it was found that 49% of Americans judged a product's quality by its warranty, 49% by availability of service, and 48% by ease of repair. So, when the company is addressing their warranty, they need to take these elements into consideration.

In developing the product warranty, the company needs to recognize that this is the same as a legal contract. Although the warranty itself may not become an issue in product liability as it relates to manufacturing negligence, it logically would become an issue in a civil suit that deals with contract law.

Government Guidelines

The Magnuson-Moss Federal Warranty Act, enacted in 1975, requires that consumers be given a complete and understandable explanation of the product warranty before a product is purchased. This isn't to say that businesses are required to provide a written warranty, because they're not, but if a company decides to offer a written warranty it must comply with all aspects of the Act. The Federal Warranty Act does not apply to services or to products sold to other companies for commercial purposes. The Act pertains only to warranties on consumer products.

Warrantors need to be aware of their legal obligations to fully inform consumers of the warranty coverage and to follow through on warranty promises. This is where the term *breach of warranty* comes into play. The wording of a warranty must therefore be well thought out. Also, the types of warranties must be known, such as:

Express warranty: A written or unwritten promise or commitment made regarding the product or service that could have been communicated verbally, or through manuals, advertising materials, or through other forms of written or oral means, which may have become the basis for the sale.

Implied warranty: A promise or agreement that may not be specifically written or stated. Is based on the common law principle of "fair value for money spent," which implies that the product is fit for a specific use or purpose.

The penalties for breach of warranty can be significant. Under the Federal Warranty Act a business can be held liable for failure to honor their warranty and be prosecuted either through a civil lawsuit—consumers can even join together and bring about a class-action lawsuit—or through government action. In addition, the courts can award damages beyond the initial costs associated with the repair or replacement of the product, and can require businesses or manufacturers to pay legal costs as well.

Full or Limited Warranties

In addition to understanding implied or express warranties, the Federal Warranty Act also requires that all written warranties for products costing $10 or more take one of three forms: full warranty, limited warranty, or a combination of the two. Full and limited warranties are defined in the Act as follows:

Full Warranty

Defective products will be repaired or replaced at no cost to the consumer, including removal and reinstallation when necessary.
Repair or replacement will be done within a reasonable time after the defect has been reported.
No unreasonable burdens will be placed on the consumer in order to obtain warranty service. (The FTC or local authorities can help interpret what is "unreasonable.") Any duties imposed on the consumer must be conspicuously disclosed.
Warranty coverage is not contingent on returning a registration card.
Warranty coverage extends to subsequent owners of the product.
Defective products will be replaced, or money refunded, if they cannot be repaired after a reasonable number of attempts.
Implied warranties may not be disclaimed or limited to the duration of the full warranty.

A full warranty does not mean:

All product parts are covered by a full warranty.
The consumer can abuse or misuse the product and still receive coverage.
The warrantor will pay for consequential damages such as towing costs, car rental, food spoilage, etc.

The product is warranted in all geographical areas (such as outside the United States).
The warranty must last for a specific time.

Limited Warranty

The warrantor places limitations on warranty coverage and must disclose them. It does not necessarily mean less overall coverage; a product may have both limited and full warranties for different parts or aspects of performance.

Some examples of possible limitations on warranties include:

- Parts only, customer pays labor
- Labor only, customer pays parts
- Customer must return a heavy product to store, factory, or service center for repairs
- Customer must pay handling or transportation charge
- Only pro-rata or credit refund allowed
- Warranty is nontransferrable.

Limited warranties may require the return of a registration card only if the warrantor does not service products if the card has not been returned.

Implied warranties may not be disclaimed, but may be limited to the duration of the limited warranty provided that the time is not unconscionably short or in violation of state law.

A limited warranty does not necessarily mean:

The product is inferior, will not work as promised, or is not as good as products covered by a full warranty.
That only part of the product is warranted.
That the warranty must last for any specified period of time.

That a purchaser can have in-warranty service done at only a limited number of locations.

After considering the legal requirements of both types of warranty, it is quite common for manufacturers to select the limited warranty as the one they will provide. As noted, however, this requires that the manufacturer clearly spell out the limitations of the warranty as it relates to the product and its performance. So, in writing out the limited Warranty, the management team needs to give thought to each aspect of what the company will cover, for what period of time, and how the return or replacement is to be handled by the customer. For instance, does the warranty cover replacement of the product, or replacement of the parts, and not any reimbursement for labor? Does the manufacturer require that the customer contact them first before returning any products or pursuing service that he or she will expect reimbursement for? If the product is warranted for parts only, does the manufacturer allow for C.O.D. returns? Also, if the product is repaired by the manufacturer, what is the warranty on the repair or replacement? If the original warranty was for 1 year, and the item required repair of some sort 8 months into the warranty, does it now have another full year of warranty or just the remainder of the original year?

Wording of the Warranty

The full or limited warranties should answer these questions:

What is covered
What is not covered
When does coverage go into effect
How long the coverage lasts
What the manufacturer will do to rectify situations
What the user has to do to get service

Especially as it relates to manufacturers shipping vast quantities of products to customers who in turn warehouse the packaged product until it is used, the question might arise of what the warranty is for product that has arrived damaged. If a package was not opened for 6 months, and then the product was found to have incurred shipping damage, does the manufacturer stand behind the replacement? Because the manufacturer would probably pursue a damage claim with the carrier, they would want to ensure that the product was inspected for damage upon arrival before going into storage. Therefore the limited warranty may stipulate that the customer is expected to inspect the contents of the product within 15 days of arrival and make the manufacturer aware of any damage incurred; otherwise the customer accepts responsibility for damage.

Another issue, especially when a manufacturer is supplying a product to another commercial user, is whether a warranty's specified duration of coverage begins on the date of manufacture (which would be identified on the product), from the date of shipment (which would have to be tracked through paperwork), or from the time received. A manufacturer may state generally that the product has a 1-year warranty from the date of manufacture, but then the product might sit in the manufacturer's own warehouse for 6–9 months before it is finally shipped or sold to the end user, leaving only 3–6 months on the warranty.

Once the Liability Expert along with the Management Team have clearly written out the full or limited warranty, they should have the policy reviewed by an attorney to ensure that it is worded properly. Warranties are governed by state and federal laws and regulations, which can be subject to interpretation. Furthermore, the attorney would want to include other legal statements or disclaimers regarding any other warranties that may have been implied or expressed. The warranty should then be printed on clearly identifiable forms that may be packaged with each product and distributed to every sales employee so that they are all fully knowledgeable of the content.

If the company or manufacturer also deal with service contracts, they would be separate from the product's warranty contract. Typically the customer or end user pays extra for a service contract, which might even be sold long after the sale of the original product. The Magnuson-Moss Warranty Act includes very broad provisions governing service contracts. For instance, if you do make a service contract available, the Act requires that you list all terms and conditions in simple language. The company offering a service contract is prohibited under the Act from disclaiming or limiting implied warranties. However, sellers of consumer products that sell service contracts merely as agents or service companies, and do not themselves extend written warranties, can disclaim implied warranties on the products they sell.

Monitoring Activity

Once a manufacturing company has its warranty well documented and in place, they should then begin statistically monitoring warranty activity in order to evaluate its effectiveness. For instance, on a monthly basis Customer Service could summarize all the calls received from customers or end users of the product, and break them down by nature of the call. If the company finds that most of the calls deal with elements of a product's warranty that was unclear, the warranty forms could be revised so as to clarify the point.

If, on the other hand, the majority of calls center on a specific problem the customers were experiencing in using the product, the company could take the steps necessary to rectify the problem with the product. The Customer Service department's monthly summary of all warranty calls received would enable the rest of the organization to identify and recognize the product issues that need to be addressed to effectively improve quality and reduce customer complaints. Such a report might show the monthly breakdown as a

Pareto chart (in a descending order of importance), which is also commonly known as the 80/20 principle (80% of the total problems are attributable to 20% of the listed causes).

In a hypothetical situation, Customer Service completes a report for each customer call received that deals with a product problem. At the end of the month, all the reports are summarized and the values of the defective products are added up and charted according to the various reasons. The breakdown may look like this:

Scratches	$600
Mounting plate problems	$75
Defective finish	$954
Nonfunctional electrically	$5000
Electrical shorts	$150
Overheating	$200
Damage in shipment	$6500
Wrong color shipped	$100
Switch malfunction	$300
Motor failure	$420

By itself this looks like merely a lot of causes and numbers. But when it is displayed as a Pareto curve, as shown in Figure 1, it presents a much different picture. You instantly recognize the Vital Few that need to be addressed versus the Trivial Many that aren't worth the immediate attention, or that wouldn't offer as much return for the effort. This example is quite simple, but take a situation where there are 40–50 Customer Service people receiving 30–50 calls each per day. What are these calls about? What is being done to address the chronic issues? How are they even being identified? It is easy to ignore the big picture and just address the smaller issues one by one, day after day. Maybe the issue to be focused on isn't a breakdown of problems as used in the example. Maybe the first issue is: What product are we receiving the most calls about? And then, why? Or, going beyond products, maybe the calls deal with other is-

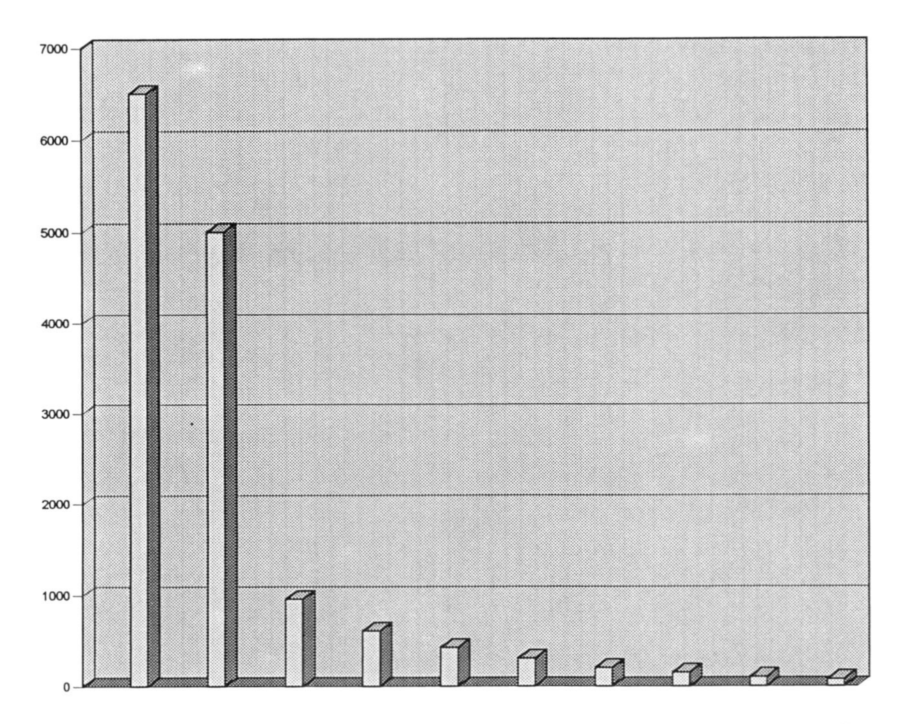

Figure 1 A Pareto curve separating the "vital few" from the "trivial many," and therefore should be addressed first.

sues, such as late shipments, incorrect billing, poor installations, or some other service aspect. The fact is, there needs to be a periodic analysis of the nature of calls being received for any reason other than to order more product.

RECALLS

At some point a manufacturer may find itself in a situation beyond the parameters of product warranty, and faced with having to initiate a recall of a product. The decision to initiate a recall ordinarily involves a high volume of product and the acknowledgment that the product out there has a flaw and must be reworked. Maybe the customer who placed the initial order and distributed the product all over the country had been bringing the same problem to the company's attention until they identified the defect, and the manufacturer is now forced to recall the product to maintain customer goodwill.

In another scenario, maybe the company's reliability-testing function identified a real potential flaw in the product, possibly due to the wrong electronic component's being used on a circuitboard, creating the potential for starting a fire. Even though it hasn't happened in the 4 months that the 5,000 products have been out there, the potential is very real and probable, so the company decides its best to recall the product rather than take any unnecessary chances. Naturally, since the efforts here are to prevent the possibility of product liability, this is the perfect opportunity to do just that. In addition, the situation having been identified during reliability testing, if the company didn't react in such a positive manner there would be strong potential for later product liability that may lead to discovery, and a charge of negligence.

The decision as to when recall is required, as well as how it will be handled logistically, entails many variables. Obviously, the nature and seriousness of the defect are the prime considerations. Management groups can be very hesitant in deciding to recall, for maybe three reasons: the cost involved, the logistical nightmare, or reluctance to admit error. However, there are ways to deal with these concerns that lessen their threat. A recall of a small number of products—one to 50, for instance—for a problem that deals with the mechanical function or performance and isn't hazardous, and in a situation where the product can be handled and shipped UPS, can normally be easily handled by just calling the end user, explaining that the product is being recalled because of a defect that will affect the product's performance or reliability, and giving the user a return authorization.

If the company is faced with a situation where the defect can present a hazardous condition, possibly resulting in injury or property loss, or both, and the quantities are large or the product itself is big, handling a recall requires a little more coordination. Not initiating a recall when it should have been, or when it falls under the guidelines of the Consumer Product Safety Act because of these obstacles, can result in serious consequences.

As stated, one of the guidelines or laws that dictate when recall is required is the Consumer Product Safety Act (CPSA), which is controlled by the U.S. Consumer Product Safety Commission (CPSC), Washington, D.C. 20207. As described in the Act, the term *consumer product* means "any article, or component part thereof, produced or distributed for sale to a consumer to use in or around a permanent or temporary household or residence, a school, in recreation, or otherwise, or for the personal use, consumption or enjoyment of a consumer in or around a permanent or temporary household or residence, a school, in recreation or otherwise."

Section 15(b) of the CPSA requires manufacturers, importers, distributors, and retailers to report to the Commission any time they receive information that reasonably supports the conclusion that a product (1) fails to comply with a voluntary standard upon which the Commission has relied under section 9 of the CPSA or (2) creates an unreasonable risk of serious injury or death. Furthermore, section 37 requires manufacturers (including importers) of a consumer product to report to the Commission if (1) a particular model of the product is the subject of at least three civil actions filed in federal or state court, (2) each suit alleges the involvement of that model in death or grievous bodily injury, as defined in section 37(e)(1), and (3) at least three of the actions result in a final settlement involving the manufacturer or in a judgment for the plaintiff involving the manufacturer or in a judgment for the plaintiff within any one of the two-year periods specified in section 37(b). The first two-year period begins on January 1, 1991, and ends on December 31, 1992. The second two-year period starts on January 1, 1993; the third on January 1, 1995, and so forth. Manufacturers must file a report within 30 days after the settlement or judgment in the third civil action to which the section 37 reporting requirement applies.

Classifying Hazards

Whether or not a company's products fall under the definition of consumer products, there is much to be gained by understanding the CSPC guidelines and classifications. For instance, they categorize product defects into three classifications:

Class A hazard: Exists when a risk of death or grievous injury or illness is likely or very likely, or serious injury or illness is very likely.

Class B hazard: Exists when a risk of death or grievous injury or illness is not likely to occur, but is possible, or when serious injury or illness is likely, or moderate injury or illness is very likely.

Class C hazard: Exists when a risk of serious injury or illness is not likely, but is possible, or when moderate injury or illness is or is not likely, but is possible.

For companies that don't produce consumer products, this classification can still be effective in the decision-making process. For instance, a Class A hazardous condition could be when the risk of death, grievous injury, or catastrophic property damage (such as fire) is very likely to happen. Class B could be when death, grievous injury, or severe property damage is likely, or when moderate to serious injury and property damage are very likely to happen. And Class C could be when the risk of moderate injury and\or potential for property damage is possible, but not as likely to happen.

Many different, and far more complex, ways have been developed to design a matrix for determining the seriousness of a field condition and deciding whether a recall should transpire. Some systems rank the type of loss and magnitude on one axis and the chances of it happening on another. Where the two axes cross gives the analyst a numerical value on, say, a 1–10 basis, which aids in the final decision on how to react. In the end, however, the decision to recall a product will rest on the executive management, and will not be an easy one.

Handling a Recall

Almost as complex as the decision to recall a product is the question of how to do it. To merely tell everyone who has the product to ship it back would be oversimplifying most cases, especially when you're not sure they even would, or such a tactic wouldn't even be practical, or you don't even know where the product is. Depending on the size, quantity, and nature of the product, and the nature of the defective condition, the question of how to address the issue can be very challenging.

Where the end users are known, and the manufacturer plans to send a letter informing them of the need to recall or handle the product in some other manner, it is important that the manufacturer be very clear on the severity of the hazardous condition and not downplay it for fear of what it might do to their reputation. As an example, the manufacturer of an electric hand drill that sells the product to other industries is made aware of the fact that a certain number of the drills were manufactured with an unsafe electrical condition that could severely shock or electrocute the user during operation. The drill manufacturer has detailed records of all the companies the drills were sold to, and determines that the only recourse is to have them sent back.

The manufacturer decides that they will mail a recall letter to each of their customers, and include with the letter a folded carton labeled with the company's address. Now they need to decide what the letter should say. If the manufacturer downplays the event in an effort to save its reputation as a quality company and elects to state that a defective condition may exist in a few of the products and they therefore request that customers return the drills in the enclosed carton for quick inspection, after which they will immediately be returned, probably only a few drills will be returned. In addition, during the weeks to follow, a small number of those who decided to keep them because of the lack of any real expressed hazard could end up being seriously injured, or even killed, and suing the manufacturer for producing a defective product. Although the manufacturer may try to claim that they made a reasonable attempt to recall their product and these customers chose to ignore the recall letter, attorneys for the plaintiffs will be quick to point out to the court that the recall letter failed to communicate the seriousness of the hazard, and therefore the defendant was negligent in their failure to warn or, technically, failed in their continuing duty to warn.

This scenario is very possible. Any company pursuing a recall that has little or no experience in the area is likely to find that a good number of their customers won't respond to the recall, regardless of how easy the manufacturer tries to make it for them. In some cases, if the product seems to be performing in a satisfactory manner at the time of the recall, the customers may decide to ignore it. For this, and other legal reasons, it is imperative that the manufacturer make it very clear that this is a genuinely serious situation.

Developing a Recall Letter

For instance, to attempt to ensure that the recall letter isn't mistakenly discarded as junk mail, the manufacturer should stamp on the envelope: IMPORTANT! RECALL DIRECTIONS ENCLOSED. The letter itself should state boldly across the top, DANGER! THE ELECTRIC DRILL YOU PURCHASED FROM OUR COMPANY

MAY CONTAIN A DEFECTIVE CONDITION THAT COULD RESULT IN ELECTROCUTION OR SERIOUS INJURY! And then the message:

> Enclosed is a preaddressed carton that we would like you to use in returning your drill to us for inspection. We will make every effort to inspect and correct the condition if found, and immediately ship it back to you. We apologize for any inconvenience this may cause, but we emphasize the seriousness of this hazard.

This is the proper way to write a recall letter in such a serious situation. The letter should be sent as certified mail. If some customers still choose not to respond, the manufacturer would be in a much better position to prove to the court at a later date that a good and reasonable effort was made to warn the end user and recall the product.

In cases where the immediate location of the product is not known, ads could be taken out in trade publications that the end users would be likely to read or posters could be placed in locations where the product is commonly marketed, warning of the impending dangers and advising the owners how to make the necessary arrangements to send the product back. The important element in a recall effort is making the hazard known and not trying to cover it up.

Coordinating Field Corrections

Arranging for a product to be returned is one way of handling a recall, but it isn't the only one. In some cases, it may not be logical or cost-effective to arrange for the product to be sent back. In other cases, the primary customer who originally ordered the product and then distributed it may not even want to return it—but he may still hold the manufacturer responsible for correcting the defect. In many of these scenarios, the manufacturer is somehow required to inspect or rework the product in the field, and figure out how to do it. If the product is still in bulk at a warehouse location, the manufacturer can send representatives to the location to perform the inspection or rework. Or the manufacturer can send one representative to the location who can then arrange for the hire of temporary employees to do the rework, which the representative can supervise.

Still another way to handle the situation, especially when the product isn't at one or two primary sites but instead has been distributed to maybe hundreds of locations, is to hire a temporary inspection agency to do the rework. As an example, say that the Royal George Department Store chain purchased 1,200 cash registers from the ABC Cash Register Co. About 2 months later, ABC recognizes that it has a defective condition in the cash registers that takes only about 20 minutes to inspect and correct, but the units are in 120 locations across the country (10 per store). The defective internal component is worth about $5. ABC knows that it would cost $60 per unit to have the cash register shipped back and another $60 to return it. And this doesn't take into consideration the added labor for shipping, repair, and reshipping, or the amount of additional loss due to the potential for the product to be damaged in shipment.

ABC communicates the defective condition to the Royal George people in the hope that they will be sympathetic to the situation and agree to help out, but Royal George is not agreeable to returning the units. Furthermore, they're not agreeable to having their own people spend any time inspecting the cash registers, and they expect ABC to correct the condition immediately.

If ABC were to send one of its own people to each of the locations, the traveling expenses (with airfare) would amount to about $800 per site. The total for the 120 locations would amount to around $96,000, and it would take about 4 months to complete the work. The Royal George people aren't concerned about the cost because they know the work is going to be done at ABC's expense, but they won't accept ABC's estimate of the amount of time it will take. They insist that the situation be corrected within 2–3 weeks. To ABC's management group, this seems impossible unless they send out crews of workers as opposed to one to three employees. And ABC doesn't have the manpower to send out without shutting down their own operation. So how can they handle this?

One way ABC can get the job done is to retain the services of a national service company, furnish them with the instructions on what to do along with the locations of the cash registers, and let them handle the inspection and rework. The expenses for this might run $100 per hour, which might amount to $400 per location × 120 locations, or $48,000. This is now half as much as it would have cost ABC to do it themselves and, because the service company has outlets in each of the locations, the work could be wrapped up in a week. Or, taking this one step further, instead of hiring the service company ABC uses a national inspection service composed of 10,000 retired Quality Assurance people living in every possible location in the country. The inspection agency handles the job for $30 per hour, which means that ABC gets the job done for Royal George within 1 week, and it ends up costing only about $15,000. This situation has now gone from a 4-month, $96,000 disaster to a 1-week, $15,000 phenomenal response.

Many companies don't realize the variety of ways possible to handle a crisis situation, primarily because they don't think about it until the crisis happens, and then no one is thinking clearly. The situation or customer demands that the manufacturer react instantly, and in doing so they may follow the most expensive course of action. This is another area that the Liability Expert can assist with, helping the company to develop a recall emergency plan or procedure before there actually is one.

Just as in a product liability situation, when a company has to immediately swing into action and initiate a recall, only one person should be piloting the process. Deciding who should do this, and how it should be done, should be done in advance. Once again, however, because the situation is so closely related to his function, the Product Liability Expert is a prime candidate. In developing the plan, the organization needs to recognize the nature of the different types of disasters it may face with its particular product line. After outlining these hypothetical scenarios, the recall expert can then investigate all the options that might be open to them, and the organization can decide on the most feasible plan to follow, and document it.

Developing a recall plan, or a number of plans based on varying potential situations, is something most companies don't think about until it is too late. Afterward, their 20/20 hindsight says that the crisis could have been handled in several other ways that would have been far less expensive and had far better results. But this has all been forgotten when the next recall disaster surfaces 5 years later and half of the players are no longer there.

12

Training the Employees

When a corporation initiates a major effort to focus on the various aspects of product liability prevention, it has to involve all the employees. Selecting someone to accept the responsibilities of the Product Liability Expert is one major step in the right direction, but the corporation also needs to create a new focus and awareness with the rest of the employees to ensure the success of the program. The employees and the various departments need to begin understanding how they fit into this new picture and effort. They must also be trained as to their roles in preventing and handling potential product liability instances and concerns.

SCHEDULING AWARENESS SESSIONS

One of the best ways to initiate this new awareness, once the Product Liability Expert has been named and the Product Liabilities Review Board has been selected and has had the opportunity to meet a few times, is to schedule introductory awareness sessions to kick off this new effort and focus. The corporate executive staff, along with the newly appointed Product Liability Expert and the Product Liabilities Review Board, can hold one or more one-hour sessions to explain the corporation's new focus and their intentions with respect to other departments.

This first introductory session should explain how serious this concern is, and the magnitude of it on a national level. To demonstrate this, many of the charts and statistics from the Preface and Chapter 1 could be used. The company may even decide to share with the employees some of its own experiences that may have been kept confidential to this point, although they may not want to get too specific in this. In addition, the CEO may want to explain to the employees the roles that will be played by the Liability Expert and each of the Product Liabilities Review Board members.

Once that has been accomplished, the Review Board itself may want to explain to the employees the legal terms associated with product liability, such as defective design, failure to warn, failure to adequately warn, reasonably foreseeable risks of harm, and reasonably foreseeable misuse. Everyone within the organization should become familiar with some of these more common legal terms and elements of strict liability and manufacturing negligence. As the Liability Expert and the Product Safety Review Board learn more about various elements of product liability and the potential exposure to product liability, they should share this new information with the other employees and departments, much as in other programs, such as Quality and Process Improvement. Although some individuals are more focused on it than others, it should not be treated as a privileged area of knowledge, or someone's own turf. A company will never be successful at preventing errors and hazards and improving products and processes without every employee's becoming something of an rank amateur in the field. At this point the Board should share with the employees what this whole concern is all about, along with their intentions as a team.

The truth of the matter is that in all probability most employees know very little about product liability, the statistics surrounding this legal area, and the specific legal terms and definitions. Of course, everyone has heard of product liability and knows from a general perspective what it is basically about: if a company manufactures a defective product that injures someone, or leads to property damage or loss, the situation will likely lead to a lawsuit. But that is about the extent of their knowledge. The rest, they feel, is up to the attorneys. So when this new Product Liability Expert—who is really one of them—pursues and gains much of this outside knowledge, at least enough to become knowledgeable in the field, and shares this knowledge with the rest of the team, he not only can talk to them on their own level but can relate what he now knows to the products they are producing. When the Liability Expert describes the terms commonly used in the legal system and shares what else he has found out—however limited it may be—the audience will usually be quite interested and willing to learn from it because, in all honesty, it is an interesting and exciting field.

From this initial awareness, or kickoff session, the group should now begin specialized focus sessions, for instance, with the Sales employees, with Customer Service, with Engineering, and with Manufacturing management. In each of these individual focus sessions, the Product Liabilities Review Board shares with the groups what they should specifically be focusing on, and about the various practices they should and should not follow to prevent the exposure to product liability.

SESSIONS

Customer Service Focus Session

With Customer Service, it must be recognized that they may be one of the first departments to actually become aware of a potential product liability incident before it actually develops. Normally, if a product fails in the field, the customer is likely to first contact Customer Service. Without the proper training, the Customer Service representative could negligently or absent-mindedly help a product liability incident to develop by either saying the wrong things over the phone or by failing to pass the communication on to others who need the information. The following hypothetical phone conversation between a customer and a representative of a major manufacturer of electric hair-curling irons is an example. The Customer Service representative is receiving 50–75 calls a day, from all kinds of customers for all kinds of reasons, when this call comes in:

Customer Service: Hello, this is Customer Service, can I help you?

Customer: Yes, I have one of your curling irons, and when I was using it the other day with my hair wet, I wound up receiving quite a shock.

Customer Service: Well, I'm very sorry to hear that. This is about the fourth time I've heard about this happening in the past 2 months. Do you still have the sales receipt and warranty, and would you like to return it?

Customer: Yes, I probably will, but why would this have happened? Don't you check these things out before you ship them?

Customer Service: Yes, we do check them out, but we make tens of thousands of those curling irons and many other products, and with those kinds of volumes there's always the potential for defective products to get out. Manufacturing and Quality Control aren't always catching these things like they should. But I do personally want to make sure that this gets taken care of for you, so if you could send that unit back to my attention, I'll make sure that it gets replaced for you right away.

Customer: Well, okay. What was your name again?
Customer Service: [gives name]
Customer: Okay. Thank you. Bye.

In the Customer Service focus group presentation (possibly changing the scenario so it reflects the company's own product line), the Liability Expert may ask, "What did you think of that? How do you think the Customer Service representative handled that call? The representative was sympathetic, wanted to exchange the product right away. Even expressed interest in wanting to handle the project personally for the customer. What do you think?" Most of the Customer Service representatives sitting there may not think the call was handled too badly, maybe wouldn't have said a few things here or there, but in general they realize that with all the calls they do receive they could have said about the same.

Then the Liability Expert really astonishes the group. "Well, let me tell you who that caller really was. That was a customer who received more than a little shock from our curling iron. That was a customer who received a serious shock from the unit, may have even been knocked out and rushed to the hospital. It is a customer who now thinks that they suffered permanent neurological damage. This phone call was a setup. The customer had an attorney present who wrote the entire script. As a matter of fact, they may have even recorded this entire conversation. They're going to sue our company for $3 million and, as part of their case, they are going to subpoena this Customer Service representative to appear at the trial, and under oath make this person expound on everything they stated in this phone conversation: 'fourth time I've heard of something like this happening in 2 months' . . . 'we make tens of thousands of these, and with those kinds of volumes defects like this can get out there' . . . 'Manufacturing and Quality Control aren't catching these things like they should.' And our representative will have to explain what they meant by all this, and they can't back out of it, because they'll be under oath."

When something like this can be presented to such a group, they can easily be scared straight. They recognize just how easy it would be to say the wrong thing to the wrong party, thinking they're merely engaging in small talk or in a conversation in which they feel they have befriended the customer calling and so can take on the attitude of "Manufacturing and Quality Control may not be doing their jobs, but I'm surely going to take care of you and do mine." It is all too easy, and can become so damaging. In another scenario, the call may even be not from a customer but from an inquisitive attorney who is preparing to launch a lawsuit and is still trying to gather some facts regarding the product and the manufacturer's practices before actually doing so, knowing that information will be hard to get afterward.

In their positions, the Customer Service people need to know how to instantly recognize the nature of calls and properly handle them. If a caller asks for some unusual technical background information regarding a product or a product's capabilities, or asks unusual questions about the company itself, the company representative should immediately ask the individual to identify him- or herself and to provide a phone number. If the caller is up to no good and wants to remain anonymous, in all likelihood he will hang up. If the caller begins to describe an accident involving the company's product, the Customer Service representative should learn to immediately recognize the potentially serious nature of the call and, rather than volunteer information, should instantly begin to collect it.

The Product Liability Expert should develop for the company a form to be used by all employees who first become aware of a potential liability incident. An example of such a form is shown in Figure 1. When a Customer Service representative is speaking to a customer who describes an accident

NOTICE OF POTENTIAL LIABILITY INCIDENT

Reported by:________________________ Date:____________

Name of contact reporting incident to you:____

Who is this contact:__________________________ Phone No.(s)____________________

____

Where is this contact located:____

Other parties who may have knowledge of the accident or incident (if known):

____________________	____________________	____________________
Name	*Position*	*Phone*
____________________	____________________	____________________
Name	*Position*	*Phone*

Product Description:____________________________ Customer:_____________________

Means by which incident was communicated:

___Phone Call
___Conversation in person
___Letter or fax (attached)
___Other ____________

Type of incident being reported:____
(Personal injury, fire, property damage, etc)

Date of incident:________________ Location:________________________________

Details as presented: ____

Please forward this form to the Product Liability Expert

Figure 1 Notice of potential product liability incident.

or injury, the representative should begin to document all the pertinent information on this form. By immediately swinging into action and becoming the questioner, as opposed to being the one questioned, the representative won't volunteer any potentially dangerous information, and would immediately deflect any inquisitive caller.

The other aspect of doing this that makes it so important is the fact that at this early stage the company's representative may be able to obtain far more information regarding the incident and surrounding details than what may be obtainable later once the caller has had time to think about it or to consult with an attorney. And the details at this early stage could be very significant for the defense.

Take the original call as an example, and let's repeat it after the Customer Service person has received the new training.

Customer Service: This is Customer Service, can I help you?

Customer: Yes, I have one of your curling irons, and when I was using it the other day with my hair wet, I wound up receiving quite a shock.

Customer Service: I'm very sorry to hear about that, and I'm also very concerned. I would like to ask you for some information regarding what happened so that I can forward to the proper people within our company, and they will be getting back in touch with you.

At this point, an inquisitive or misrepresented caller would probably end the call right here. If it was a legitimate call, the Customer Service representative could begin to collect vital information regarding the situation without placing the company in any jeopardy. For instance, after asking the caller for his or her name, address, and phone number, the representative could ask for details regarding the accident or injury, as well as the extent of damage or injury. The caller may have been stunned by the shock in this case—and maybe mad about what happened, and wants the company to know it—but right now just feels that she is entitled to a new unit and, if there were any additional costs incurred, may feel that she deserves restitution. Maybe the caller hasn't discussed this situation with anyone else, and knows and volunteers that a doctor has pronounced that there was no serious injury; she was just a little shaken up by the experience. If the company exchanged the product, and maybe paid for any immediate injury or damage, she would be satisfied. So the caller gives all the details to the Customer Service representative as requested, and waits for the company to respond.

Now, let's say that this person talks with a few friends or fellow workers, and some of them recommend that the individual talk to an attorney and consider suing the company. The attorney indicates that there may be a pretty good case and instructs the individual not to talk about the case with anyone from this point on (the attorney may or may not know of the conversation that has already taken place). When the company's Product Liability Expert calls back, the individual states that she has hired an attorney and has been instructed not to talk with anyone from now on. The manufacturer is now shut off from any more information. But the Customer Service representative has already received a tremendous amount of information from the customer, including pertinent information regarding how the accident happened and, more importantly, the extent of the injuries, including the fact that the customer stated that there weren't any permanent injuries. This kind of information will be very helpful for the defense—and very troublesome for the plaintiff, when they later imply that the injuries are serious and presumed to be permanent.

But suppose the incident doesn't follow that course and the customer just waits patiently for someone from the company to call back. The Liability Expert and the Review Board decide to have the same Customer Service representative, or the Manager of Customer Service, phone the customer and ask that the defective unit be returned to his personal attention, along with any other associated bills the customer incurred (say, an emergency-room-visit bill for $200), and promises to send the customer a new unit right away as well as take care of the bill for them. The customer feels very good about the service received and pursues the case no further. The company prevents a potential liability case and is out only a few hundred dollars at most.

In this situation, we will assume that the company knew that their protect must have malfunctioned, maybe because they had already seen others that had done the same. This isn't to imply that the company would necessarily always take this course of action. A caller might, for instance, claim that the product shorted out and started a fire, causing $3,200 in property damage. The company wouldn't just jump to reimburse this customer unless it was an unusual situation and they knew once again that there was a problem with the product. Instead, they might ask the customer to return the product for inspection, with the customer hoping for subrogation. The manufacturer would then analyze the product and report to the customer that the product was not defective, nor was it the cause of the fire: it was just caught up in the fire, which must have started somewhere else. The customer would logically have to accept this determination, whether or not he agrees with the analysis, because, first, he no longer has the product, and second, the product has now been substantially altered. Therefore it would now be unlikely that an attorney would want to take on the case.

In both situations, the manufacturer not only wards off a lawsuit but has also learned something. By getting involved early and having the opportunity to analyze a defective product, or through testing and analysis determine that the product wasn't defective, it becomes a learning experience that could help the Expert to become an even greater asset to the company.

But the key point in all of this is the importance of developing a form for reporting potential liability incidents and getting anyone who may have contact with customers to use it whenever they are made aware of such an incident, whether actual or potential.

Engineering Focus Session

As the Product Liabilities Review Board conducts their focus session with the Engineering group, they will focus on other aspects of potential liability that are pertinent to them, such as recognizing potential dangerous aspects of certain product designs, the necessity to develop adequate warnings and instructions, and reviewing once again the legal terminology and how it pertains to their product.

To start, a product can technically be defective and considered unreasonably dangerous just because the engineers or the company failed to adequately warn the end user about a hazard that is inherent to the design. The engineers need to accept that not all product knowledge is common knowledge, especially as it relates to what might be considered unequal knowledge, or knowledge beyond what a reasonably prudent person might possess. So this is one of the elements to share in the engineering focus session: recognizing that it is essential to identify any risk of harm or foreseeable risk of harm that is inherent in the product, and adequately warning against it.

Here is one consideration in developing and determining the placement of a warning:

> The determination of whether a warning is adequate depends upon a balancing of considerations including, among other factors, the severity of the danger, the likelihood that the warning will catch the attention of those who will foreseeably use the product and convey the nature of the danger to them, the intensity and form of the warning, and the cost of improving the strength or mode of the warning [Bloxom v. Bloxom 512 So. 2d 839 LA. 1987].

Here is another perspective on that same element:

> Even though a product may not be defectively designed so as to be dangerous to one who properly uses it, a duty to warn may exist if the manufacture has reason to believe a user or operator of it might so use it as to increase the risk of injury, particularly if the manufacturer has no reason to believe that the user will comprehend the risk [407 N.W. 2d 92 Minn. 1987].

So, in general, there are many new elements that the Product Liabilities Review Board is asking the engineering group to consider when designing a new product, and in developing the associated warnings and instructions. Four initial guiding principles or questions for the engineering group to focus on when critically analyzing their new product are:

Could the product be viewed as unreasonably dangerous for the marketplace?
Could it be determined that there are insufficient or inadequate warnings?
Could there be a reasonably foreseeable risk of harm that the company is failing to warn against?
Is there an element of reasonably foreseeable misuse that isn't being addressed?

These are some of the new thoughts and concerns that the Review Board would want to share with the engineering group in their focus session. It isn't that the Board or the Expert is expected to have all the answers at this point, but as the specific product issues are now presented to the engineering group and they become unsure about how to address them, a group consensus may be reached with the engineers along with the Board, or the engineers along with the Expert. A key element that needs to be kept in mind by all parties at this point is that they do not need to become so concerned now about product liability that for every question that arises the company needs to consult with an attorney for the absolute legal answer. That is not what is being proposed. Let's keep in mind that up to this point most organizations had been somewhat oblivious to this whole issue, and yet they were still in existence. So, as with any Quality program startup, the company isn't expected to start out as experts; it is just expected to start out.

Manufacturing Focus Session

The next group that the Product Liabilities Review Board may want to schedule a focus session with is the manufacturing group. In addition to the original awareness presentation charts and statistics, it may be beneficial for this group to also see some elements of what was presented to Customer Service and Engineering. As an example, if the manufacturing focus group includes Manufacturing Engineering, Process Engineering, Quality Assurance, and other specialized departments in addition to the standard manufacturing departments, they all act as backup and added assurance that the company as a whole is critiquing the product and its engineering for the same critical elements that Engineering is looking at.

Beyond these initial elements, the Board's focus would be to stress the importance of ensuring that the proposed product is adequately and routinely tested, and that the key aspects of assembly and construction that could add to the potential for product liability are somehow being monitored or controlled. Keep in mind that much of the emphasis up to this point has been on the evaluation of the proposed design and adequate warnings, but not the potential for negligence in manufacture by building the product contrary to specification. This is where that element has to be brought into the picture. In emphasizing the importance and the magnitude of product liability throughout the country, and the new focus on prevention being initiated by the corporation, the Board in this focus session wants to create a new interest and effort in the control of what is being produced. Here are some examples from other cases of injuries due to defective product conditions that might possibly have been caught during production, and the amounts awarded by the courts:

A drill bit shattered during use because of an excess amount of sulfur in the steel, necessitating removal of the plaintiff's eye. $125,000

Excess vitamin D added to dairy products caused hypercalcemia in one plaintiff and the death of another. $500,000

Plaintiff sustained back injury when thrown across room from the shock given by a floor-maintenance machine. $85,000

Death of several crew members on a plane due to defective instrument approach charts. $12,785,580

An automobile fell on plaintiff due to defective bumper jack, causing death. $570,550

An artificial respirator malfunctioned due to selector-valve design defect. $4,000,000

An industrial mixing machine caused death when hopper door became stuck. $2,000,000

Plaintiff lost four fingers while operating a press with no safety devices. $83,800

A pressure gauge failed to indicate air-pressure buildup. The explosion caused extensive injuries. $155,829

A U-clip securing scaffolding suspension cable gave way. $4,500,000

From these diverse cases, we can see why each department needs to recognize its specific role in product liability prevention. Engineering deals with safe product design, warnings, and instructions, but Manufacturing needs to focus on the potential for defective product or product made contrary to specification.

Besides the tests to ensure that various electronic and gauge components are working properly, there is also the mechanical function test of other aspects of the design. It is amazing how assembly personnel can live with a defective aspect of a design without its ever being corrected. They bring a condition to the attention of their supervisor, who in turn makes Engineering or some other department aware of it, and both parties go back to what they were doing without addressing the defective condition. The assembly personnel become frustrated and ignore the situation from that point on. Doors or mechanical parts don't fit or interlock properly because they are dimensionally inaccurate, or because the stack-up of various parts affects the overall dimension, and the final assembly person forces the fit to get the product through assembly.

In another situation an electrical component is functioning intermittently; when it does function, the assembler passes the product through as opposed to investigating the possible cause of the intermittent condition. Maybe 150 or 1,500 of the products will pass through the assembly workstation that day, and the employee may not have, or may not want to take, the time to bring it to someone's attention. So the next person who will be exposed to the defective condition and product will be the ultimate customer.

The manufacturing management and the factory employees need to realize the importance of inspecting their material and work, following specifications, and reacting to instances of variance and other defective conditions. Just like management, they need to be shown the magnitude of product liability and the importance of their efforts in the prevention of it. It is very common for manufacturing personnel to be completely oblivious to the customers' experience with the products they made. The factory employees continue to manufacture product day in and day out. When product is found to be defective in the field, the issues are addressed and resolved by management and the workers are never made aware of them. Yet in most cases the workers could have the most impact in controlling future recurrences.

13

Customer and Supplier Agreements

In manufacturing or subcontracted service work, one area of the product liability prevention effort that becomes crucial is contractual agreements. Normally companies are good at making sure everyone is aware of their invoicing and payment terms and conditions, and they are pretty good at having something written regarding their warranties, but when it comes to liabilities, the contracts are often vague, outdated, or don't even deal with the subject matter.

For many companies, incidents of product liability don't surface very often, so it wouldn't be a topic of great concern. The language in contracts or purchase orders is limited and sometimes antiquated. The rest of the fine print on contracts and purchase orders is periodically updated, but the section on liability tends to be carried over from the old form because nobody knows who wrote it, or what it should even say. So it goes unchanged.

Then one day the sleeping company is hit by a catastrophic accident in the field and everyone is searching for and reading all the fine print they ever published or received, to determine how well they had protected themselves or what their liabilities may be to others. But it is all too late at this point. What's written is written, and the case goes forward. Customers and suppliers who once had a pretty close working relationship may now become estranged, or even enemies. In the end, maybe the parties involved learn something from the event or maybe they don't. Or maybe they make the necessary changes and improvements only to those contracts or order forms that apply to the one situation, until the day they are hit by another catastrophic event that affects another area. The corporation needs to review and update all its contractual agreements, sales orders, purchase orders, practices, and procedures that do or should address the area of liabilities. And the best way of doing this is with the newly developed Product Liability Review Board.

The Board, or specific members of the Board, can begin to meet with each of the departments within the corporation in individual focus sessions to review the existing contractual agreements and begin to make the necessary improvements and revisions. The Board will seek recommendations from their insurance representatives, the in-house attorneys for the insurance carrier, and their own corporate attorneys on agreements and contractual language. It is not to be expected that the Board or the Liability Expert have the knowledge necessary to solely handle these areas of concern; this is really a legal affair and requires the guidance of a legal expert. This chapter identifies areas of concern and recommendations on how certain concerns should be addressed, but it will not specifically state how contracts should be worded because of the unique aspects of most companies.

There is no single place where this focused effort must start, because within a short period of time all areas should be covered. And almost any area has as much opportunity for problems as another so, depending on the nature of the business, the Review Board should begin with the areas of greatest relative exposure and work downward. This chapter identifies various areas one by one, with no degree of importance being implied.

CUSTOMER AGREEMENTS AND PURCHASE ORDERS

Especially in job shop environments, where you are the supplier and you are receiving purchase orders for product from customers, contractual agreements with customers and the language in purchase orders are matters of importance and need to be reviewed. Such fine print is often overlooked and assumed to state "standard terms"; more important are the order itself and how fast the company is going to be paid for it. Customers normally expect that the supplier or manufacturer of a product will indemnify them in the event of a product liability incident, but their contracts and purchase orders may not reflect that requirement or expectation. This then becomes a "customer goodwill" problem if such a situation happens, because the insurance company handling the manufacturer's risk coverage isn't automatically going to indemnify anyone unless contractually required to do so.

For instance, a sports arena hires a manufacturing company to design and build a large scoreboard to be hung above their main stage, displaying not only the score but promotional elements, advertisements, and the like. The arena issues the manufacturer a purchase order, which in fine print on the back explains that the arena won't pay for the product until it is completely installed to their satisfaction, that the manufacturer is expected to warranty their work, and, if the product is going to be shipped late or any other complications arise, that the supplier is responsible for advising the customer. The job shop manufacturer designs and builds the board, ships it to the arena, and hires a crane company to install it.

Six months later, the scoreboard breaks loose from its suspension system and crashes to the floor, injuring a dozen spectators. Attorneys for the injured parties immediately initiate lawsuits against the arena and the scoreboard manufacturer. Soon after the initial actions have been started, the attorneys find out that a third-party company actually installed the board and name them as codefendants. All three parties are now being sued and have to begin their defense work. The arena, however, immediately contacts the manufacturer through either their insurance carrier or their attorneys, and demands that the manufacturer indemnify them in this case since they had no part in the scoreboard design, manufacture, or installation. Furthermore, the arena's owners, who are building similar facilities in other major cities and have been working with the manufacturer on design concepts for those locations as well, indicate that future business with them could be in jeopardy if the manufacturer doesn't relieve them of this legal burden and expense. Realizing what may be at stake, the manufacturer makes a request to their insurance carrier that they indemnify the arena, thereby technically handling the arena's defense as well as the manufacturer's. The carrier reviews the manufacturer's policy to see if the arena is listed as an additionally insured party, which it isn't, and asks the manufacturer whether there is any contractual agreement with the arena to indemnify them. The manufacturer reads all the copy on the back of the purchase order they received from the arena, along with any other papers or agreements, and doesn't find anything referencing such a requirement. The carrier declines the manufacturer's request, because of the added expense they will incur, and now the manufacturer is in a bit of a predicament. Telling the arena owners that they cannot indemnify them even though they agree that the fault will logically lie with either them or the installation company is going to create a lot of tension and ill feeling. But they cannot demand that the insurance carrier indemnify the arena owners, especially when such provisions were never previously agreed to.

This is a delicate situation and, unfortunately, even though the supplier or manufacturer is sitting on the hot seat, the customer himself is partially to blame because of his failure to incorporate the proper language into the agreements and purchase orders. A manufacturer can add primary customers to their "additionally insured" lists, and in that way supply the necessary protection for such incidents. It isn't practical for a supplier or manufacturer to review a customer's purchase order and advise them to add such wording. And there is a concern from the other perspective that shouldn't be overlooked, to assure indemnity when it may be found in fact that the customer was at fault for the incident. So not only should there be such clauses in contracts, but the clauses must specify indemnification only in situations where the supplier is found to be solely responsible for the incident.

Although the fine print in a purchase order agreement is commonly the vehicle used for communicating such concerns and requirements, more and more companies are adding supplements to these purchase orders that more specifically, and in greater detail, elaborate on what the customer requires of the supplier. These addenda may entail another page or two of requirements that the supplier or manufacturer is expected to comply with.

SUBCONTRACTOR REQUIREMENTS AND PURCHASE ORDERS

Just as the ultimate customer or end user should ensure that contractual agreements with manufacturers to supply specific products are comprehensive, so should the manufacturer assure such protection with its own subcontracted companies. A subcontractor in this context is another company that the primary manufacturer has contracted either to produce a finished product for them that will be shipped directly into the field or to perform a field service for them. In both cases, the final customer or end user will hold the primary manufacturer responsible for all product and services supplied, regardless of who actually performed the work. But whether the manufacturer is holding such subcontracted parties equally responsible is another question.

Prior to initiating such an alliance with an outside company, the primary manufacturer needs to ensure that the subcontractor has adequate liability insurance. For some small subcontractors, this may be beyond their financial abilities, or it may not have ever been a prerequisite up to this point. Regardless of how small the service to be performed seems, there may still be just as much potential for an incident of liability as there could be if the job were many times the size. So, part of the approval process for new subcontractors should be that they need to provide a certificate of insurance, and it must be for a designated minimum amount of coverage, such as $1–5 million.

It isn't uncommon to hire an independent engineering contractor who can perform a necessary service for a very competitive fee, and who is well recognized in his field, only to find that the individual does not carry any liability insurance. The contractor does protect himself, however, by incorporating himself and listing practically nothing as an asset. This may never be known to his clients, and may never even become an issue, unless there is an incident. Then the independent contractor quickly folds up and the manufacturer faces the lawsuit all by themselves.

Another scenario is that the subcontracting company carries an inadequate amount of liability coverage, such as $250–500,000. It should be kept in mind when considering the amount of coverage needed that it is very common for plaintiffs to request $1–$3.5 million for personal injury and punitive damage. In such a case, $250,000 of coverage would obviously be inadequate. Counsel for the plaintiff would then pursue the party with the deepest pockets: the manufacturer. In addition to requiring that the subcontracted party have adequate insurance coverage, the manufacturer must also make sure that they are being indemnified. This can be achieved as part of either their upfront agreement and approval process or the fine print in the purchase orders issued.

Beyond the insurance coverage, the primary manufacturer needs to secure certain agreements with the supplier or service company to ensure that they are protected in this area as well as others. The primary manufacturer must ensure that there is an indemnification clause in some contract, purchase order, or agreement that clearly states that the subcontractor will indemnify the primary manufacturer in any situation that leads to a claim or litigation where the subcontractor is at fault.

In addition to liability, there should also be a clause that refers to the subcontractor's warranting their work. Most large manufacturers find themselves paying for additional expenses related to warranty work, as well as customer concessions. They have efforts in place to try to control such expenditures, mainly as part of their Quality program, but it still happens and can easily involve thousands of dollars monthly, depending on the manufacturer's size. However, they can also find themselves paying the expenses for correcting something that should be the responsibility of the subcontracted company. For instance, a primary manufacturer subcontracts the production of a laser printer to another assembly house because the primary manufacturer is behind in production on other items. The subcontractor quotes a price to assemble and ship the units directly to the primary manufacturer's customers. The primary manufacturer offers a one-year warranty on the units to their customers.

Some of the units break down in the field after 60–90 days. The primary manufacturer's Customer Service personnel could ship replacements to the customers, honoring their own warranty on the product, and never even go after the subcontracted assembly house for subrogation. Or, even worse, Customer Service could have the defective units sent back to the subcontract location for repair and work out a "repair cost" with them, separate from the original contracted assembly cost. In another words, the primary manufacturer pays the subcontractor to build the product originally and then, because they did a poor job of it, pays them additional money to repair the product. It isn't hard for such a scenario to transpire if no one is paying attention.

Or it could be that the subcontractor performed a field-service function on a product, such as installing a product in the field or servicing a product that was malfunctioning. A month or two later, the product fails again or the installation proves to be defective, and the primary manufacturer sends the same company, or another company, back to the location to reservice the product. If the primary manufacturer is going to offer the end user a one-year warranty on the product, then they need to equally ensure that the subcontracted company is standing behind any work they do for the same period of time. This is especially true with field-service and installation companies; they are right there on the spot and not only can see what the problem is that they need to repair, but are also able to see its cause, and should rectify that as well. Otherwise, they know that the product will supply them ongoing business at the manufacturer's expense.

As an example, the primary manufacturer may have built a central air-conditioning system for a small office building. They offer a one-year parts and service warranty on the unit. Two months after it is installed, it suddenly quits. The manufacturer contracts a local service company to go and inspect the unit and determine why it isn't working. Upon inspection, the service contractor finds that water leaked into the unit and knocked out an electrical circuit. The serviceman replaces the electrical circuit and sends the manufacturer an invoice, showing the labor to replace a blown circuit. A month later, more water leaks in and the circuit is blown again. And again the serviceman goes on call and replaces the circuit and bills the manufacturer. The manufacturer isn't there, so he has no idea why this electrical circuit keeps blowing. But the serviceman knows exactly why.

With the new contractual agreement in place, the manufacturer contracts the service company and asks them to service the central air system, but stipulates in their contract and purchase order that the subcontracted company must furnish a one-year warranty on their work. Should the unit become defective during that time frame and demonstrate the same effects or end result, the subcontracted service company will be expected to reservice the unit free of charge. Naturally, with this in mind, the serviceman not only replaces the electrical circuit but addresses the water leakage as well.

In some instances a larger manufacturer may find that when they explain these new requirements to subcontract companies the subcontractors will tell them that they can't financially back such agreements, that they are too small. In essence, they are willing to do $3,500 worth of service work for you or assemble, ship, and bill you for $1 million worth of product, but they don't want to accept any responsibility for it. This is the type of company that primary manufacturers need to steer clear of, even though they quite probably are the lowest bidders on the job. Companies fail to add up just how much such companies are really costing them, because they see only small individual invoices and fail to ever summarize the big picture.

SUPPLIER AGREEMENTS AND PURCHASE ORDERS

A supplier, in this context, is a vendor, catalog company, or job shop that is supplying the primary manufacturing company with a component part to be used in the manufacturer's final product. The component may be a stock shelf item or a component custom-made by a job shop to the primary manufacturer's specifications. Controlling liabilities and product warranties in these scenarios becomes a little trickier. Even if a manufacturer issues to all these types of vendors purchase orders that incorporate an indemnity clause, enforcing such liabilities could be another question. When a manufacturer orders stock components from other suppliers or vendors, those companies typically have no idea what the manufacturer intends to do with them and therefore cannot logically be held responsible for some of the consequences. If a component part catastrophically fails, the primary manufacturer will in all probability be the only company sued, for the plaintiff won't logically know the source of the component part, if they even know that the situation was the result of some component.

A manufacturer has two basic alternatives if it is known that an accident or injury was in fact caused solely by a component and was not the result of an engineering error or error in application. First, the manufacturer, upon entering into a liability action, can make it known to the other party that the incident was due solely to the failure of the component, and was not the fault of the manufacturer or their designed product. With this knowledge, the other party may address their actions toward that supplier. But if the incident has entered litigation, in all probability the plaintiff will just name the component manufacturer as a codefendant in the action. The manufacturer can at that point ask the supplier to indemnify them, and legally hold them harmless in this action. If the supplier agrees, and if the plaintiff and court agree, the manufacturer could then be released.

The other alternative is for the manufacturer to demand subrogation from the supplier for any and all legal costs and losses incurred if the manufacturer handles the action on their own, or even to be reimbursed for expenses if they are codefendants. A surprise could be in store for the manufacturer if it is found that the supplier doesn't have liability insurance or the financial assets to pursue defense, and ends up folding because of the action. In such a case, the manufacturer may end up being the sole loser, with little recourse.

If the component was supplied by a job shop and built to the manufacturer's own specifications, or designed by the supplier for the manufacturer's specific application, then the manufacturer may be held responsible for failure due to it technically being their own product. This type of a supplier is a borderline subcontractor and should be treated as one. For instance, in entering into an agreement to have the supplier furnish the specific component, the manufacturer should qualify the supplier. In doing this, just as in a subcontracted arrangement, the manufacturer should first ascertain that the selected supplier has adequate liability insurance. Second, the manufacturer should have a written agreement assuring that the supplier fully intends to stand behind their product and all warranty costs that are incurred.

14

Accident Investigation

Beyond all the efforts to prevent product liability, the Quality system, the focus on developing an internal expert, the Design Reviews, reliability testing, and labels and instructions, getting involved in the actual accident investigation is the other major focus of this book. It is a significant departure from the way a corporation typically handles an incident in the field, but it is the ultimate opportunity. Everything else mentioned above needs to be in place before initiating this activity, but this is where a corporation can make a significant difference in its future.

A corporation's getting involved in its own investigation and defense, much less taking a major role, is a new concept for American industry. Many periodicals have resisted even publishing articles promoting this practice, because of their uncertainty and fear of dabbling in legal areas and the possible repercussions. This role and responsibility had been reserved solely for the attorneys and insurance people up to this point, but that is all beginning to change. And the ironic aspect of all this is that defense attorneys themselves encourage such involvement and participation. The vast majority of defense attorneys do not view this as sacred ground, and would like to be able to work with the corporation in question or its designated Liability Expert. But, since they are normally retained by the carrier and have little or no initial contact with the client, they tend to feel that initiating such interaction would be frowned on, especially by the party who retained them in the first place.

As much as the legal defense would normally welcome such corporate involvement, they themselves don't become involved until the situation escalates to that level, which is part of what we are trying to avoid. The first party the corporation has to work with is the insurance carrier, which will initially be a little more difficult. It isn't so much that the insurance industry doesn't want to have the corporation involved in the investigation (although it may not); it is more often found that they are used to doing everything themselves.

It is the insurance company's habit to fully handle and investigate any incident that the corporation makes them aware of, and this is where the difficulty lies. The Product Liability Expert needs to break into this exclusive arrangement and get the insurance company to begin viewing them as an equal partner, if not the director, in any activity planned. The only advantage the corporation has from the start is that they will logically be the first to ever know of any situation.

BEGINNING THE INVESTIGATION

When a company first hears of a situation in the field that resulted in an injury or property damage, they need to investigate it as quickly as possible. The situation may have been brought to the organization's attention by the victim himself, a third party, the victim's own insurance carrier, or an outside attorney.

Depending on the severity of the incident, if it was reported by the victim or a third party the Product Liability Expert may elect to just call and get the details. For instance, if the company manufactures a wall-mounted display case or a coffeemaker, and the report from the field was that the case fell to the floor or the coffeemaker malfunctioned and poured boiling water all over the counter and floor, the Liability Expert could determine that the extent of the damages would be contained in the hundreds of dollars and therefore it wouldn't be worthwhile to travel to the location to investigate unless it was nearby. In these kinds of cases, the customer may hold the company liable for his loss, but it probably wouldn't escalate into a lawsuit. The company in turn would most likely reimburse the user for his loss. The Liability Expert would still head some type of investigation back at the company into probable cause and corrective action.

ARRANGING THE INITIAL MEETING WITH INSURANCE REPRESENTATIVES

But let's say that the coffeemaker caught fire and burned part of a room or building. Now the situation has a higher level of severity, and the Liability Expert should make every effort to get there as soon as possible. In this scenario, the situation may still have been brought to the company's attention by the victim, a third party, or even the victim's insurance company, which is looking for subrogation. The Liability Expert should contact the company's own insurance carrier to make them aware of the incident, informing them that he would like to get there as soon as possible, and asking them who they might want to be present representing them. Although the company's Liability Expert will always be the same person coming from the same location, most insurance carriers will send representatives who are nearest to the incident. This may take a few days for them to decide, but hopefully it won't take very long because fire damage isn't normally left standing very long. And if it was a commercial establishment, they will want to get it cleaned up and back in business as soon as possible.

When the Liability Expert first goes out on such an investigation, he must recognize that it is first and foremost a fact-finding mission. The purpose is to see the product and hear what happened. If the product at the investigation is found to be defective by the Liability Expert, the company would want to react. If the product wasn't defective, the Expert may still want to react in some manner.

In this scenario, with only the insurance companies from both sides present, the Liability Expert would first want to view and photograph the remains of the actual product. The objective here is to determine if possible whether the product appeared to be defective or manufactured in a defective manner. Depending on the complexity and nature of the product, he may have to use other types of gauges or instruments for this analysis. If the Expert is able to determine anything, one way or the other, nothing should be said at this point, nor should he explain or elaborate on any testing or readings being taken that may make the outcome obvious to the observer.

Once the Expert has viewed the product, explicit notes should be taken regarding any and every aspect of the product in its current condition without noting any specific defects. This may be the last time the product is ever available, or it may be noticed at some later date that the product was altered between this meeting and the next. It is therefore essential to have thorough notes and photographs of how the product appeared at this point in time.

The next objective is to learn everything possible about the actions that led up to the accident. It is hard to list here the kinds of questions that need to be asked, given the diversity of products that we could be talking about, but the

Expert needs to act like a police officer investigating an accident scene and gather all the facts available. Using the coffeemaker as an example, here would be a typical list of questions that he would ask the user's insurance company:

How did the end user receive the coffeemaker?
How long did the user have the unit prior to the incident?
How many hours per day was it operating?
What time of the day did this happen?
Who was present?
Had the unit ever shown any signs of malfunctioning previous to this?
Where was the coffeemaker located?
What actually happened?
Were any reports written by anybody from the scene of the fire (fire reports, insurance company investigation reports)?

The Expert should ask every question that would be relevant to the product and incident to fully understand the events that led up to the incident. He may need to revisit these notes a year from now, so he shouldn't try to commit what is being said to memory; it must be written down. Specific details that may not appear important at this point may turn out to be significant at a later date, so everything said should be documented.

Once the Expert has viewed the remains of the product and heard all the details, he may be in a position to recognize whether the product truly malfunctioned or was designed or manufactured in a defective manner. If that is the case, he would not make any comment or indication implying this, but would state something along the lines that they will take this information back to analyze, and will be getting back to all the other parties shortly—in other words, doing whatever it takes to politely excuse himself and leave. Afterward, he can consult with the corporation and the insurance carrier to best determine how they should handle the case and any compensation.

If, on the other hand, the Expert is of the opinion that the product was not defective, nor did it malfunction, he would want to take this opportunity to express this expert opinion. This is a very critical moment in the investigation, and the Expert has to be very meticulous about everything he says. This is also one of the primary messages of this book: take the best advantage of this exact moment and try to put an end to this potential product liability action.

If the Expert is convinced that it wasn't the product that was at fault here, but the user or some other contributing factor, he wants to explain this to all the insurance parties present, in the most convincing manner possible. He will in all probability be viewed as the most knowledgeable party present about the product's capabilities. So, if this expert opinion is presented properly, it should persuade others, and the effort here will be a success.

For instance, let's say hypothetically that the insurance carrier for the other party theorized that the hotplate overheated and caused ignition to the surrounding plastic, which led to the burning of the rest of the room. Maybe they based this on stories heard about similar situations with other coffeemakers. If the Expert explains to the group that the thermistor (or whatever is used to regulate the temperature) on this unit was the proper one and is still reading accurately, that everything was wired properly, and that the surrounding plastic was flame-retardant so it couldn't have started a fire, that would begin to cast doubt on the theory and may convince them to look elsewhere. And the Expert may even see other possibilities of how the fire started. Even though that isn't the objective here, if there is another strong possibility that steers the parties in another direction, it may dissuade them from the coffeemaker theory.

Having this involvement and opportunity is a major new inroad for any manufacturing corporation. As you can see, the technical information presented regarding the product would never have been known by anyone else who would logically have been present under normal conditions, so the theory presented may have been the theory believed. You can also see that the Expert played a major role in asking all the right questions, which contained technical inference about the performance of the product.

The Expert at this point wants to come across as being very concerned and very helpful, but demonstrating or projecting self-confidence in what he says, in a convincing manner that isn't offensive. The other parties will never have the amount of technical knowledge that the Expert possesses of the specific product or know its true capabilities along these lines, and they will recognize that at this meeting. So, they are likely to be persuaded by what is being said, if the Expert handles this in a convincing and sincere manner and doesn't appear to be talking down to them or insulting their intelligence.

It is common for individuals (plaintiffs and their insurance representatives) to say things at these initial informal meetings that they regret later, or that future counsel would have advised them against. But asking the right questions at the time regarding the events and documenting all the answers will create a valuable historical document.

When the Expert returns from this investigation, he will want to gather all the information obtained and put it together in a final report that may be distributed to others within the organization. The key point here is that the final report just states the facts as they were presented and is void of the Expert's opinions, even though they were discussed at the meeting. If this incident doesn't die in the weeks and months to come, the Expert will need to revisit these notes to remember the condition of the product as it appeared at the investigation, and recall the facts as they were presented at the meeting.

But the notes themselves are not legally protected in any manner. This is why any written reports need to be void of the Expert's opinion. In this scenario, where the only parties present besides the Expert were the various insurance representatives, all reports written regarding the incident could eventually fall into the hands of attorneys as both sides prepare for legal battle. Because many of the minor elements presented could be forgotten in the months to come, having written them down will be a definite advantage.

If, however, the report contains the Expert's comments and opinions as to why the manufacturer isn't at fault, this could prove later to be a fatal error. This is basically the foundation on which counsel for the defense would eventually build their case. In our scenario where the Expert explained why the product wasn't defective and what may possibly have caused the fire, if the other insurance company wasn't convinced by what was said and decided to pursue this through the legal system, the first thing their attorney would do is initiate a request for documents along with an interrogatory and require from the manufacturer any reports written regarding the incident. Once counsel had the Expert's report with his comments and opinions, they would go out and find their own expert to evaluate and disprove the findings. Then it would be a situation of one expert fighting another expert. And if counsel for the plaintiff could discredit something that was written, it would tend to discredit the manufacturer's in-house Expert. Chances are that when this investigative meeting took place, the other insurance company didn't take many notes regarding the Expert's opinion on why the product couldn't have been defective, but was just listening intensely. So when they decide months later to retain an attorney to fight the case, the attorney will be at an initial disadvantage and have to try to gather all the facts that the Expert already has. The opinions once expressed by the Expert will now become part of the legal defense strategy.

If the case were to escalate to this stage, not only would the opposing insurance company be retaining counsel, but the manufacturer's risk carrier would be doing the same. As the opposition requires weeks or months to gather their initial data, the Expert can now arrange a meeting with the chosen defense counsel. At this point the manufacturer's own insurance carrier will tend to begin bowing out and leave the handling of the case with the attorney. This is all the more reason why it is beneficial for the corporation to remain actively involved. Without this expert backup, counsel for the defense would be no better off than plaintiff's counsel in clearly understanding the case.

In this transition, however, the Expert can make immediate contact with the selected attorney, and completely lay out the case along with the findings. This is extremely helpful to the new attorney, and eliminates hours and days that would otherwise have been spent by the attorney in gathering all the pertinent information. This is also one of the reasons that attorneys tend to be big promoters of the movement toward companies' involvement in incident investigation.

In addition to bringing the new attorney up to speed on the case and the technical findings, the Expert can now work with him or her on future defense strategies. If they want to initiate their own depositions or interrogatories, or depose the plaintiff or other witnesses, they can begin planning for this. Furthermore, although the Expert would have shared with the attorney the initial report, the Expert's opinions and backup information can also be documented now and forwarded to the attorney and be protected from discovery, because it would fall under the category of attorney–client privileged information—as information asked for by counsel for the defense, it isn't discoverable by the plaintiff.

MEETING WITH FORENSICS EXPERTS

In some situations, such as those that involve fires, the insurance company pursuing subrogation may have the Product Liability Expert meet with them at a forensics laboratory with their own expert witness. Normally attorneys aren't involved in situations like this; initially it may be just the other insurance carrier's representative and the forensics investigator, or it could even be an arranged meeting between the Expert and the forensics investigator. They may have a theory about the cause, but tampering with the evidence is required to prove out the theory.

In this type of scenario, normally the insurance companies are looking for both experts to analyze the product and agree on whether it was defective. It may be recognized that the product will have to be opened up, or tested and analyzed by a methods that would alter its original condition, and therefore both parties need to be present. The product or component would be photographed by both parties, and they would agree on how to proceed with the analysis. Once the product has been thoroughly analyzed, both experts would attempt to agree on the results, which they would be communicated back to the insurance companies involved.

In another situation, both insurance companies may also be present at the meeting, and the opposing insurance company wants their expert to fully explain why the product was found to be defective. They are convinced that the expert knows what he is talking about, so the objective here is to convince everyone else. Handling this type of situation may be a little more challenging for the Expert. Of course, if the product really was defective, there may not be much to be said. But if the Expert disagrees with the forensics expert's evaluation and theory, it could lead into a lot of discussion dealing with some facts, a lot of theory, and some speculation. On one hand, the forensics expert will undoubtedly have a substantial list of credentials and want to prevail in front of their client. On the other hand, the forensics expert will make the same amount of money either way, and so may not be overly concerned.

Although the forensics expert will not have the technical knowledge regarding the product that the company's Product Liability Expert will likely have, there is a distinct possibility that he or she has more technical knowledge in the field. For instance, the company's Expert may have 10 years of working knowledge with the coffeemaker, but the forensics expert may have 10 years of knowledge and education in electronics and other related products, and be able to discuss characteristics that push the Liability Expert to the limits of his own knowledge.

In this scenario, being an ambassador of goodwill may take a back seat to getting the technical point across. Hopefully, the two expert parties will agree on the outcome, or at least the Liability Expert will be far more convincing to all parties than the forensics expert. But if the company Expert feels that the forensics expert is off base, he shouldn't hesitate to say so, because if the parties can't reach mutual agreement here, the opposing party is likely to initiate legal action. So the company Expert needs to be as convincing as possible—if in fact the product truly wasn't defective.

ARRANGING THE INITIAL MEETING WITH ATTORNEYS

When the Product Liability Expert is first made aware of a situation in the field via legal channels or legal correspondence as opposed to an insurance company, he will still want to arrange this initial fact-finding mission. But it would be handled a little differently. When the incident is being presented by another insurance company, or a second or third party, the Expert can contact the company's own insurance carrier, see who they want to have present for the investigation, and go. When legal parties are involved, it needs to be recognized that the incident has already risen to another level and won't be dropped or written off as easily. The Expert now needs to contact the insurance carrier and determine whether the carrier wants to immediately retain and have counsel present for the investigation, send a representative for the investigation to determine the case's potential and then decide whether to retain counsel based on the outcome of this initial meeting, or just allow the Expert to meet with the other attorney and report back on the findings.

Let's go back to our coffeemaker example, but in this new scenario it somehow resulted in an injury rather than a fire. At first it may be implied that the attorney is just looking for medical compensation along with compensation for lost earnings, but the attorney is alleging that the product was defective or that it malfunctioned. The manufacturer and its Product Liability Expert are very interested in seeing this product and hearing what transpired, and determining if their product was in fact defective. Let's say that the insurance company tells the Expert to go ahead with the initial investigation and that they don't have anyone available that they can have present as well. So they recommend that the Expert meet with the attorney, view the product, and report back to them on the findings.

The Expert would then call the attorney and set up a date and time to meet and discuss what happened. Once again, this is merely a fact-finding mission, and the Expert is just interested in having the opportunity to inspect the product and learn the particulars. If the attorney hasn't initiated any litigation, he or she won't have any problem with this. For the Expert, however, this is substantially different from the meeting with insurance representatives. Insurance companies in these cases are normally fully prepared to cover the losses their clients experienced, but are just looking for subrogation from the manufacturer if the product really was at fault. The attorney is also looking for compensation for his client feeling that the product was in fact defective, and that's all he's looking for.

When the Product Liability Expert meets with the attorney, he has to handle this situation in an entirely different manner. The Expert needs to examine the product and ask all the same questions he did in the meeting with the insurance reps. The difference in this meeting is that the attorney here is going to be substantially more interested in any of the Expert's findings. In all probability the attorney won't know very much about the product at this early stage. If any other expert analyzed the product, it was probably a brief analysis. The attorney will therefore be very interested in any tests or analyses made of the product during this meeting, in the effort to gain more knowledge.

When they first meet, it is important for the Expert to begin to do a character analysis of the attorney. The expression "the first 10 seconds can tell you a lot about a person" really applies here. Understand the type of person you're dealing with. Note the personality, the openness with which you were received, the attitude toward this incident and this meeting. All the same objectives still apply to this meeting: view the product, get the facts, and try to correct any erroneous conclusions. But it has to be handled very delicately and diplomatically.

The attorney is going to take many notes at this meeting. He will note anything the Expert finds or comments on in the analysis of the product, and anything else said. As the Expert asks questions about the events leading up to the incident, the attorney will be vague. If the Expert finds that the product is not defective, he will want to explain this to the attorney but should keep a perspective on the type of person he is dealing with. For instance, the attorney may have a really bad attitude toward this whole thing, and not be very receptive to even having this meeting. If the Expert says anything that upsets him, he could instantly walk out of the meeting and handle everything through the courts from that point on. The Expert doesn't want this. If he knows that the product isn't defective, and feels that the user caused the accident or injury through their own negligence, he wants to communicate this in a very sensitive manner. In all likelihood, the attorney doesn't fully understand the product and its capabilities at this point, and may even have some doubts about the client's account of what happened. If handled effectively by the Expert, this will be a cordial meeting and make the attorney rethink the credibility of the whole case. If handled improperly, he may never agree to meet with the Expert again and will pursue a lawsuit.

In addition to writing down anything the Expert finds upon analysis, and the opinions expressed, the attorney may employ another tactic that at times can be very effective: saying little or nothing, even while appearing very friendly and receptive. This can make the Expert feel uneasy and, to maintain a friendly atmosphere, he may begin to ramble. And this rambling could be very dangerous.

To appreciate the effect of this, think about a situation where you had to travel in the car with an important customer whom you had never previously met. You have nothing in common that you know of, only your business. It is important for you and the company you're representing to have a positive impression on this customer, so you want to maintain a friendly and sincere conversation. But let's say the customer is smiling but not doing much talking. Two minutes of silence can seem like an eternity, so you desperately search for more to say. And the more nervous you become, the more you tend to ramble and the more likely you are to say something you will regret afterward.

In this situation too, the only thing you both have in common is the product. You're not going anywhere, so the subject matter for discussion is very limited. In this uneasiness, things could be said that one would regret later, only it would be too late because the other attorney would have written it down. For instance, say that the attorney thought that the coffeemaker had a defective switch or connection to the switch. The Expert tests the switch, along with several other components, and finds that everything appears to be okay (although the Expert may know that there are potential inherent problems with that specific switch, which the attorney doesn't know). Saying merely that there isn't anything wrong with this switch would discredit the attorney's original theory and caused him to reconsider pursuing this any further. But, in the silence, the Expert volunteers that although there isn't anything wrong with this switch or this unit, on future units the company has decided to use an altogether different type of switch because of potential problems. Now the Expert has said something that may eventually be regretted. In fact, it may be enough to shed new light on the case for the attorney who pursues it in earnest.

Some legal parties may caution that no one other than an attorney should handle this meeting and investigation just because of these potential slips, but that is an overreaction. The Expert just needs to develop the ability to work with other attorneys, and think carefully about everything he says. Keep the meeting and investigation professional. Friendly, but professional. Don't get so relaxed that you begin to ramble and possibly say things that you'll regret the minute they come out of your mouth.

INITIATING CORRECTIVE ACTION OR PRODUCT IMPROVEMENT

Hopefully, when the Expert carries out these initial fact-finding missions, it will be found that the product wasn't defective and he will succeed in convincing the other parties and stopping any further action in its tracks. But in some situations the Expert may recognize that there are problems. Maybe the design truly is a defective concept, the unit manufactured didn't comply with the specifications developed, or the product lacks the proper warnings or instructions.

None of these defective conditions should be acknowledged at the time of the joint meeting, but they can't be ignored either. As much as the corporation's efforts here are to get involved in product liability defense, the corporation has to put even more effort into prevention. And even though it is too late to prevent this incident, it isn't too late to prevent the next. When the Expert returns to the company, a report will be written that will just explain the facts as they were presented. The opinions will be communicated orally. But the Expert needs to spin right into action on any problems that were detected with the product, and initiate corrective action. Bringing together either the Product Review Board or the management team, the Expert needs to fill them in on the problems that were detected, and discuss how to get them corrected. This is an undocumented game plan because what the future holds regarding this specific case may be unknown, so memos regarding the incident and corrective plans should be avoided to prevent the possibility of future discovery.

Even if the product truly was innocent, in many such investigative meetings especially with attorneys, comments will be made regarding the possible lack of specific warnings or instructions. And even though the case may not be pursued any further, the Expert should react to this advice and make any improvements possible or practical.

DRIVING THE CASE

Once the Expert returns from an investigation, whether with insurance reps, forensics experts, or attorneys, he needs to stay on top of the case for the corporation. With insurance carriers and outside attorneys handling incidents and cases, they tend to drag out for months and even years, as discussed in Chapter 3. There is a tendancy by both parties to open the file about once every month or two, and initiate some type of action. And because of this, months down the road they even find themselves asking redundant questions because they don't remember where they left off.

Now that the Product Liability Expert is involved and is at times even driving the case, the Expert can keep the project moving and help bring it to an end. In the instances where the Expert met with the other parties and appeared to pretty much convince them that the product wasn't defective but that the accident or injury may have been largely the result of a negligent action, he will probably lie in wait to see whether the parties initiate any other action. But in situations where the case is likely to continue on, the Expert can be the impetus in keeping the other parties active and making things happen faster than they would have otherwise.

This assistance will also help keep expenses down for the insurance company and save the attorney time and expense. If there are aspects of the product that need to be looked into more thoroughly, the Expert can handle this for the attorney. Or if any theories need to be proved out through testing, the Expert can handle this for legal counsel, and supply full reports that would be protected as attorney–client privilege or have work-product immunity.

15

Planning the Defense

The primary efforts in product liability are focused on preventing it in the first place, or successfully stopping it in its tracks when the potential case first arises. After the initial investigation, the other party may be just another insurance company or the actual plaintiff. There still may not be any legal counsel involved, which would mean that the case either is going to die or will enter litigation. Once the investigation has concluded, the corporation will normally have submitted their findings or opinions to the other party. The ball will technically be in the other court. At this stage, however, the corporation will want to make every possible effort to discourage the other party from taking legal action. It is one of their last chances of stopping the case in its tracks by convincing the plaintiff that he or she has no case.

But if these efforts have failed, then the corporation needs to start thinking ahead and become actively involved in planning its defense strategies with assigned counsel. This is a function that up to now was handled entirely by the attorneys or in conjunction with the risk insurance carrier if the corporation was insured. But now the corporation, through their Product Liability Expert, are going to become actively involved in planning the legal strategy.

ATTORNEY SELECTION

If the corporation is self-insured, they would logically hire a good product liability attorney at this point. If the corporation is fully protected with risk insurance, the insurance company would retain counsel or assign in-house (insurance) counsel to the case. In some cases, however, the corporation may elect to retain their own counsel to represent them and pass on legal counsel that the insurance company may otherwise have sought out, or have the two counsellors work in conjunction with each other. If the insurance carrier is large enough, it may have its own in-house legal staff, which it can now deploy to begin working on this case. Furthermore, because it is in house, their actions won't be regulated or restricted. In this scenario the corporation should benefit from this legal representation, and the Liability Expert should be able to maintain a constant dialog.

If the insurance carrier retains local counsel, counsel will in all probability not have free rein. The attorney will probably have to submit an estimate of costs, along with an idea of their chances of winning. This is a very common situation, and can to some extent handicap representation for the defense. In this scenario, the corporation may elect to retain their own local (to the company) attorney, one who specializes in product liability cases. This attorney would probably be used in all future cases as well, which would give him or her a good working relationship with the corporation, and an eventual good working knowledge of the company's product lines. If this is the case, the corporation would make the carrier aware of this decision and the carrier could decide whether they still want to retain their own counsel or totally depend on the corporation's attorney.

If, however, the corporation does go with the insurance company's locally retained attorney, the Liability Expert may find that he or she has to drive the project much of the time. Trying to get a preliminary investigation meeting planned by the defense attorney with plaintiff's attorney could take a month or more. The same could be true with many other motions, inquiries, and actions. It seems that most outside attorneys have too many projects going on simultaneously, and it takes continuous follow-up to ensure that things are getting done. If this begins to be a chronic problem with the insurance company's retained attorney, the Liability Expert should contact the insurance representative that the company primarily deals with to complain about the slow service, and perhaps even recommend that they retain new counsel.

PLANNING DEFENSE STRATEGIES

Once this has been decided, and the Liability Expert knows who the counsel will be that he or she will be working with, they can begin working on the case. After having performed the initial incident investigation and heard the facts surrounding the case, the corporation as well as counsel for the defense want to plan their strategies. Since this is the defense, in all probability counsel for the plaintiff is already pretty active, and may have submitted interrogatories along with requests for documents. Of course, the first strategy planning meeting should have taken place in concert with the initial investigation, or the first follow-up meeting with the plaintiff that the defense attorney is present at. This is where the Expert can begin to discuss the product itself, the possibility of the alleged cause and result, his own theories of probability, and other aspects surrounding the incident.

From this first meeting with defense counsel, the Product Liability Expert wants to maintain a constant path of communication so that both parties remain in sync with each other and continue to discuss various aspects of the case, new thoughts and considerations, and proposed action plans. This needs to be a very close working relationship, with few if any other parties becoming involved. This means that the Product Liability Expert should be the only one from the corporation who is communicating with counsel for the defense, as opposed to any other members of the Product Liabilities Review Board.

In product failure cases, the Liability Expert may want to conduct various failure-analysis studies on samples of the product line to determine whether the product could have failed as suggested. This is also where other members of the Board could come into play, especially if the testing and analysis functions are under their direction. At this point, and in cooperation with counsel, the Expert can perform these tests under the premise of work-product immunity. Performing these types of tests is a definite advantage for the defense.

In many situations, the case up to this point has been built around speculation and theory on the part of counsel for the plaintiff, their expert witnesses, or the expert witness the other party's insurance company may have hired if it is a situation where they are looking for subrogation. In any case, the defense is going to be in a much better position than would counsel for the plaintiff to test other product and determine the credibility of the accusation. Counsel for the plaintiff is normally looking at the failed unit in question, or may have been successful in obtaining samples through other means. Still, the product will logically be foreign to them and their experts, and there are sure to be many unknowns.

TYPICAL PLAINTIFF STRATEGIES

In the initial complaint filed in court, the plaintiff may have implicated several parties who may have been connected with the product in some manner. This could include sales or distribution people, component suppliers if they are known, the primary company identified (if the product was made for a specific customer whose name is being advertised), as well as the manufacturer of the product. This is common practice for an attorney initiating a lawsuit on behalf of a client, because they aren't sure who the real defendant should be, and how all the rest are related. By initiating the action against everyone possible, you're sure to have included the guilty party if in fact someone really is negligent, and the facts can be sorted out later.

Furthermore, with three to five parties named in the complaint, there is also the chance that they will pursue a settlement and divide the loss among themselves. For example, if the attorney initiating the action is ultimately pursuing $1 million, rather than try to get that from one defendant, maybe each of the five parties involved would be willing to absorb a $200,000 loss. Or maybe three or four of the defendants would like to bail out early and settle for $200,000 losses, while the corporate player decides to ride it out to the end. That is also a legitimate alternative, and technically the plaintiff scores a victory to a large degree, without even trying.

PARTIES NAMED IN THE COMPLAINT

In any case, the first action, as stated, is to implicate everyone possible in the lawsuit, making them all codefendants. This makes it difficult for the corporation who really manufactured the product and who is primarily going to defend it, because they will have a hard time controlling the case.

The rest of the attorneys involved are going to create a lot of confusion; they're all going to have their own opinions on how it should be handled and they will all need to be present during meetings, investigations, depositions, and any other collective meeting. There will also be more paperwork for the corporate Liability Expert. In all probability, the Expert will be receiving interrogatories to complete not only from the plaintiff's attorney but from all the codefendant attorneys as well.

As stated earlier, two of the standard complaints that are initiated by counsel for the plaintiff in a product liability action are lack of adequate warning and that the product is unreasonably dangerous. With either of these allegations, counsel can initiate the court action without much effort or research, which also translates into less up-front expense and investment. So what we have is a successfully launched lawsuit with established grounds, a number of parties implicated, and an attorney who is hoping that the parties will settle.

An attorney who does put more effort and study into the product and design may feel that he or she can pursue the case from a defective-design perspective. With this approach, the attorney will consult with experts to determine what state of the art is with respect to the product design, and what alternative designs may have been. Counsel for the plaintiff hopes to prove that alternative designs would have been safer and were feasible for the manufacturer to incorporate, but that they failed to do so. In addition to consulting with outside experts on the design, the attorney will also hope to discover information from the manufacturer that will show that such alternative designs were considered but rejected, possibly for economic reasons or, better yet, after risk–benefit analysis. Any information later obtained along these lines would also demonstrate the defendant's knowledge of potentially dangerous propensities inherent in the product.

THE INTERROGATORY

If there isn't any positive immediate reaction on the part of the defendants to pursue an early settlement—especially the primary defendant, who is the manufacturer in a case with multiple defendants—the next step for the plaintiff is to begin the process of gaining more knowledge about the product and the manufacturer. This will initially be accomplished through the use of the interrogatory, a questionnaire developed by either attorney that is served on the opposing party through the court system in an effort to gain more knowledge and detail regarding any related aspects of the case or background on the other party.

If the attorney submitting the interrogatory isn't very knowledgeable about manufacturing, or doesn't have a strong product liability background, the interrogatory may be a standard form of basic questions. But if the attorney is an experienced product liability trial lawyer, the interrogatory can be quite lengthy and very comprehensive, at times to the extent of seeming overwhelming. Counsel for the plaintiff will have the interrogatory delivered to the defense attorney, who will in turn forward it to the client—specifically, the Liability Expert. This will normally be the second step or phase in the litigation process, with the first step being the official complaint or notification of the lawsuit.

When the interrogatory is sent to the Liability Expert from the defense attorney, it normally has to be completed within a certain time frame, such as 30 days. This is ordered by the court. The person who completes the interrogatory is also under oath and has to clearly identify himself and sign the last page. The answers must be truthful or the respondent could be subject to legal consequences. Normally, in addition to the Expert, the defense attorney is also preparing a response to the questions of an interrogatory. If the Expert is concerned about answering certain questions, he can consult with the defense attorney. In most cases, the defense attorney will just tell the Expert to answer the questions to the best of his ability, and the attorney will then review the answers prior to returning it. By doing this, the defense attorney can object to certain questions asked, or submit other answers that are far less incriminating or informative, and thereby protect the defense. A sample of an interrogatory is offered below, following the definition of terms.

Definitions

Defendant refers to the company and, where applicable, its officers, directors, employees, partners, agents, attorneys, or anyone else acting on defendant's behalf.

Product means the product alleged to have caused the plaintiff's injuries as mentioned in the plaintiff's complaint.

Incident means the circumstances and events surrounding the occurrence that gave rise to the action.

Person means any natural person or any business, legal, or governmental entity or association.

Document means any type of written communication, and includes the original or a copy of handwriting, typewriting, printing, photocopy, letters, or any other tangible thing or form of communication.

Identify, when referring to a person, means to give to the fullest extent known the person's full name, present or last known address, phone number, and current place of employment.

Identify, when referring to documents, means to supply to the fullest extent known the type of document, general subject matter, date of document, the originator and recipient.

Sample Interrogratory

1. *Identify* all persons who supplied information that was required to answer these interrogatories.

2. *Identify* the owner(s) of *defendant* company if privately held, or parent company if part of a larger corporation.
3. *Identify* each and every individual who may have had a role in the design and engineering of the subject *product.*
4. State when and where the subject *product* was designed or engineered.
5. Did this *product*, when developed or designed, constitute a replacement, improvement, or change in a prior product that was manufactured, assembled, or sold?
   A. If affirmative, identify the prior product(s).
   B. State the dates at which the prior products were developed or designed by this defendant.
   C. State the reasons why the prior product(s) were replaced, improved, or otherwise changed.
6. *Identify* each *person* who may have specific knowledge of the manufacturing process of the *product.*
7. Fully describe all design specifications relating to the *product.*
8. Describe all applicable government, independent agency, trade, or industry standards, codes, or regulations, local or national, that may pertain to or have inference for the subject *product.*
9. *Identify* any other individual, consultant, company, or laboratory that may have performed or conducted an evaluation of the product, and who may have had any input into the design and engineering of the *product.*
10. Describe any and all Quality Control steps and procedures that would have controlled the engineering and manufacturing of the *product.*
11. *Identify* any and all Quality Control *documents* that would have contained information pertaining to the manufacture of the *product.*
12. *Identify* any reliability test that may have been performed on the product, the initial design of the product or the products components, the individuals that conducted the reliability tests, along with the results and dates of such tests.

13. *Identify* any and all Quality Control personnel who would have been involved in the inspection, audit, testing or evaluation of the *product* during manufacture.
14. Describe any changes or improvements that may have taken place regarding the design or engineering of the *product* since the *product* in question was produced.
15. *Identify* any instructions or manuals that would have been supplied with the *product.*
16. Describe any changes or improvements that may have taken place involving the labels, instructions, or manuals that are affixed to or accompany the product since the *product* in question was produced.
17. State in detail specifically how the product was intended to be used, along with any other foreseeable uses.
18. *Identify* any warnings or precautions regarding potential dangers or hazardous conditions to the customer(s) or end users of the *product*, other than those affixed to the *product* itself or contained in the *product's* instructions.
19. List each month, day, and year that the product was manufactured, the quantity that was produced during each production order, and where the products were shipped for a two-year period prior to the *incident.*
20. *Identify* the month, day, and year of any instance where product was found to be defective, either through during normal inspection or from customer notification, that required inspection or rework, and include the defective condition(s) found and the extent of the rework, and *identify* the parties involved in the Quality Control inspection.
21. *Identify* any other reports of customer or field problems with the product, or products of similar nature, that had to be investigated or corrected.
22. *Identify* all individuals who have become involved in this *incident*, have detailed knowledge of the *incident*, or maintain certain facts or opinions regarding the *incident*.

23. *Identify* all individuals who are expected to testify on behalf of the defense, the nature of their testimony, their background and experience, and current positions.
24. *Identify* any and all *documents* or reports that exist regarding the specific *incident* or facts surrounding the *incident* or this case, who initiated each of the *documents* or reports, and whom each of the *documents* or reports was distributed to.
25. State the invoice number, sales order number, or other means of identification that pertained to the sale of the specific *product* involved in the *incident.*
26. Set forth in detail the facts on which you base each of your defenses, including the names and addresses of any individuals who may have contributed to these facts and opinions.
27. If you allege that the plaintiff is guilt of contributory negligence, state the facts and opinions on which you base this defense.
28. At any date after this specific *incident*, did the *defendant* become aware of other potential or actual incidents or claims regarding the same product or similar products? If so, *identify* the parties involved in any other claims or incidents, along with the date(s) of those other incidents.

As shown in the sample interrogatory, the questions can really dig into the heart of the operation, and uncover not only how the company operates but specific considerations regarding the subject product. As the Liability Expert answers these questions, it becomes a delicate situation to not make known to the plaintiff every problem experienced and at the same time not perjure oneself. This sample interrogatory is relatively short. It is quite common for a real interrogatory to be 50–100 questions long, which not only allows for in-depth investigation into the operation but is also time-consuming to answer.

The limitations on interrogatories are that they cannot be unduly burdensome, overbroad, difficult to answer, or overwhelming or oppressive in nature, or they can be objected to by defense counsel. An interrogatory might be considered overwhelming or unduly burdensome to answer purely because of the number of records and documents that the defendant would have to review in order to answer the questions. One recourse for the defendant in such cases is Option (c) of Federal Rule 33, which allows the defendant to identify in the interrogatories the specific records and documents in which the answers can be found, and then allowing the other party a reasonable opportunity to examine those records and obtain the answers themselves. An extreme example of the use of this option is found in Tytel v. Richardson-Merrell, where the information sought in the interrogatories would have required the defendant to review over 100,000 documents; the court held that the interrogatories were unduly broad and oppressive, and allowed the plaintiff to instead view for themselves microfilm records.

DISCOVERY

As if the interrogatory weren't inquisitive enough, they eventually lead to discovery. This means that, through the questions asked and answered under oath, counsel for the plaintiff can demand copies of all the records, orders, correspondence, blueprints and other specifications, and anything else that was mentioned or identified that they feel is pertinent. Counsel for the plaintiff in essence "discovers" all these new documents and records that may substantially help them with their case.

This is when we also get into the concerns regarding dangerous documents. Dangerous comments about the product may have been made not only in customer and field correspondence but also in various Quality Control reports, engineering change requests, and several other areas. When a potential product liability situation first surfaces, it can make many members of management very uneasy. In the earliest stages, there may not be any attorneys involved, nor the possibility that an immediate lawsuit is even being considered, so uneasiness about a defective product's surfacing in the field and causing problems for the customer or end user could have generated a flow of finger-pointing or other damaging correspondence that would now be regretted.

Another point to consider during the submission of documents, especially those that involve the Quality programs and procedures, is that the procedures are normally interlinked. Some procedures identify other procedures, documents, records, and so on. Eventually the other attorney has gained a pretty in-depth understanding of the whole Quality program and will demand to see all the records that are being referenced and were supposed to be kept.

TRADE SECRETS AND PROPRIETARY INFORMATION

An element of defense for consideration is the concept of trade secrets and claims of business privilege. In some cases the product in question, or the processes by which it is manufactured or the suppliers of the process or manufactured part, may be considered proprietary to the manufacturer or confidential. When a plaintiff requests such information through the interrogatory action, the defendant has the right to claim that it is a trade secret and/or confidential business information, and that disclosure of such information could be harmful to the defendant. Such information is protected under Federal Rule 26(c)(7), which allows that "a trade secret or other confidential research, development, or commercial information not be disclosed or be disclosed only in a designated way." The criteria for determining whether certain material falls under these guidelines are as follows (Restatement (First) of Torts 757):

1. The extent to which the information is known outside the business
2. The extent to which it is known by employees and others involved in the business
3. The extent of measures taken by the defendant to guard the secrecy of the information
4. The value of the information to the defendant and the defendant's competitors
5. The amount of effort or money expended by the defendant in developing the information
6. The ease or difficulty with which the information could be properly acquired or duplicated by others

Once it has been established that the information requested qualifies under these guidelines, it is up to the court to determine what will be allowed. The plaintiff will have to demonstrate to the court that the information is relative and needed or the court will prevent disclosure. If the plaintiff can demonstrate such necessity, the court may allow discovery but issue a protective order that specifies whom the information can be disseminated to.

DEPOSITIONS

Once the litigation has gone through the interrogatory and request for documents stages, it enters into the deposition stage. The deposition is like an interrogatory, only the party being deposed is questioned in person, with no time to think about it. The deposed party is also under oath, so unless an attorney objects to the question the other attorney is asking the witness has little choice but to answer the question. Although the two primary people to be deposed are the plaintiff and the Liability Expert, who is also serving as the expert witness, each party could elect to depose many others who were involved in some way with the product in order to help their case. This then leads into more evidence and discovery.

By identifying in the interrogatories the various members of the organization who were involved with the product, such as those in Quality Control, Supervision, Engineering, and Sales, counsel for the plaintiff also acquires a good list of candidates for the depositions and other possible court testimonies. Appearing for this hearing can make people very nervous. Being cross-examined by counsel for the plaintiff, compounded by the legal setting and the implied consequences, can cause many witnesses for the defense to make progressively dangerous statements that begin to bury the manufacturer.

When these other corporate personnel are called on to appear at a deposition, all fear can break loose. Appearing at a deposition hearing or in court at the trial will be extremely intimidating for most employees, especially because they haven't had any exposure to the whole process up to this point. It is not uncommon for these employees to fear that they may go to jail if they don't spill everything they've ever known about the product and management decisions and practices. This is where the Liability Expert needs to step in and help counsel the employees. For instance, it is important for anyone about to testify to know that when he is asked a question, he should only answer it specifically; he shouldn't elaborate on the answer or begin to enter other areas of concern that aren't part of the specific question.

Here's an example of what I mean. You are the Product Engineer or Quality Manager for a manufacturer of space heaters. Your company manufactures 10 models of space heater. It is alleged that one of your units malfunctioned and caught on fire, burning down an office building. Up to this point it hasn't been proven, and in fact your own testing and analysis indicate that the scenario isn't probable. However, there was a problem with one of the other models you manufactured and there were instances with that unit that resulted in fires—but not this unit. You are now testifying at a deposition and are under oath. Even though this isn't an actual trial, a transcriptionist is there, along with attorneys for both sides.

After you are sworn in and asked to identify yourself, and a few other preliminary questions, counsel for the plaintiff begins to ask you very specific questions regarding the product and the operation. You are nervous about being asked to testify, and even more nervous about saying things that either could be devastating to the company or you fear could be viewed as perjury. The attorney asks whether you have any knowledge that these specific units have ever caught fire in the past. Here is where the proverbial imperative "just answer the question" comes into play.

This product line has no history of ever catching on fire. But other product lines have. You should answer the question simply "no." But because of your nervousness, you're going to try to be evasive, and technically accurate, with your answer, so you say, "No, not this specific unit." Even though you have told the truth, you have quite obviously indicated that there were other problems. So the attorney quickly comes back with a second question, whether there have been fires with any other unit. Now you have to answer the question, and you are forced to disclose this new information that may never have been known had you just said "no."

Even worse than trying to be smart or evasive is the belief that you have to reveal everything. When the same question is asked, you answer, "No, but we did have problems with other units that we manufactured over the years." There wasn't any reason to say all that, but you felt that it was your duty to be truthful and tell everything you know, so you do just that and end up eventually crucifying your company. You feel that you had no choice because you were under oath, but you did—all you had to do was answer the specific question with a specific answer. And this could be the scenario with any of the employees who could ever be asked to testify. They can become very nervous about the event and begin to reflect on all the other problem situations they have knowledge of. The Liability Expert needs to work with those who are asked to testify and help calm their fears.

From a different perspective, having the in-depth knowledge that the Liability Expert will likely possess makes it difficult even for that individual to be the sole witness on the part of the manufacturer. For instance, if the question asked by the opposing attorney isn't whether there had ever been any other fires with the specific model but whether the company had ever experienced fires with any of their units or whether they were involved in litigation of any sort regarding their products. Most other employees wouldn't have any knowledge of other cases or incidents because they are being handled by the Liability Expert. But now the Expert is under oath and required to answer, and this too can create problems.

In situations like this that can lead to a tremendous amount of discovery, the testifying witness can consult with his or her own attorney before answering, and possibly the defense attorney can object to the question on the grounds that it isn't relative to the case at hand or on some other grounds. Or the defense attorney can enter on record the fact that they object to the question, and even though the Product Liability Expert answers the question, making known certain new revelations, a judge could review defense counsel's objection and rule that the answer can't be used by the plaintiff.

Advantages of Outside Independent Experts

Another strategy to consider is the use of an outside independent expert witness. Counsel for the defense may pursue an outside expert opinion on the possibilities of product failure or for structural analysis, in addition to what the manufacturer can offer. This has been the common practice up to this point, prior to the manufacturer's taking a lead role. In any case, the outside independent expert normally has a long list of credentials, including a substantial academic background, which adds to the individual's credibility as a testifying witness or expert.

The manufacturer's Liability Expert might consider consulting with counsel to determine if it would be a good strategy to allow the outside independent expert appear at depositions, and possibly testify in court, to help reduce the potential for discovery. The expert can answer technical questions regarding the product and certain performance aspects, comment on industry standards and regulations, and cover many other aspects, especially if the expert is coached by the company's Liability Expert.

Another possible advantage of using an outside expert witness may be the individual's credentials. Although the Liability Expert is likely to have more technical knowledge than anyone else regarding every aspect of the product, certain cross-examination by the plaintiff's attorney can make the Expert's testimony questionable. Say that the product in question is electric or electronic, and the case centers on an electrical shock or a short that led to a fire. The manufacturer's Liability Expert has a bachelor's degree in political science or journalism, or doesn't have a degree but has 15 years of experience manufacturing the product. The attorney for the plaintiff will start out the deposition or testimony by questioning the individual's educational and professional background.

The attorney would first ask the Product Liability Expert's name, position with the company, and educational background. The Expert states that he or she has a bachelor's degree from such-and-such university. The attorney asks what the degree is in, and the Expert replies political science or journalism. The attorney decides not to go on with asking about the individual's job experience, because the Expert at this point appears to not be qualified to give highly technical answers and the attorney wants to leave it that way. Meanwhile, the independent expert witness for the plaintiff has a master's degree or a Ph.D. in electronic engineering, and goes on to elaborate about his years of service as a consultant in the electronics field. The witness for the plaintiff could come off as being more credible than the Product Liability Expert.

If the defense retains an independent outside expert who can match the plaintiff's expert's credentials, then, along with the Liability Expert's assistance and playing the role of the in-house expert, they can succeed in presenting a major challenge to any opponent. However, if the defense doesn't retain an outside independent expert and continues with the Liability Expert as its primary witness, during cross-examination the defense attorney would want to ask the Liability Expert numerous questions about his professional background and technical experience, to establish his credibility as an expert on the product.

But there is a distinct advantage for the defense in using an outside independent expert: he or she will lack the internal knowledge of other incidents and litigation that could lead to discovery and problems. Surviving discovery is a major challenge for the defense. The litigation phases of interrogatories, requests for documents, and depositions are all loaded with opportunities for the plaintiff's lawyer to discover pieces of evidence that could substantially help their case and cripple the defense to such a degree that settlement is the only logical recourse. For the defense, you don't win discovery; you only hope that you can survive or withstand it.

Deposition Tactics Regarding Expert Witnesses

One of the questions both parties will undoubtedly face back in the interrogatories is to identify the expert witnesses expected to testify in the trial. It is best to state that this has not yet been decided, but eventually the names will have to be supplied. The tactic commonly used by both the plaintiff and the defense is that, once one side's experts are made known (for the defense, probably including the Liability Expert), the other party schedules a deposition of that expert.

The primary purpose of this deposition is to get the other side's expert witness to present the information and opinions he or she has regarding the product and the incident, and expound on all the formulations and calculations used to reach a conclusion. The experts will be asked to identify the assumed facts that support their theories, as well as to highlight the most significant facts. Having this detailed information prior to the trial can be very beneficial to the opposing party, because it allows them to prepare a significant challenge to the theories and calculations at the trial itself.

The defense, along with the Liabilities Expert, will be in the best position to spend a lot of time analyzing and testing various aspects of the plaintiff's expert's theory and finding errors in the calculations that would allow them to launch a formidable attack. Counsel for the plaintiff, however, may not want to spend the funds needed to prove or disprove defense's facts regarding the product's capabilities, nor would the attorney and expert witness, because of their other business demands, have the time available for such required testing and analysis.

In some cases, plaintiff's expert witness may present theory rather than scientific fact. The theory of how the product failed or malfunctioned, leading to the accident or injury, would appear to be very impressive when described by such an expert, when in fact it may be nothing more than "junk science," lacking any validity. Having identified this in the deposition stage is a major advantage for the defense. The question then becomes whether the defense should act on this knowledge before or during trial.

Debunking Plaintiff's Expert Witness

In the landmark case Daubert v. Merrell Dow Pharmaceuticals Inc., 113 S. Ct. 2786 (1993), the United States Supreme Court addressed the admissibility of expert opinion for the first time since the adoption of the federal rules of evidence. The case itself dealt with a suit brought against Merrell Dow Pharmaceuticals regarding limb-reduction birth defects affecting two minors, allegedly because their mother had once taken a drug called Bendectin for morning sickness.

The significant aspect of the case was that the theory presented by the plaintiff's experts did not reflect scientific knowledge, nor were the findings or theory derived by scientific method. The theories were presented by experts with impressive qualifications and under the assurance that their conclusions were reliable. This basis for such testimony is referred to as the Frye test, which originated in the case of Frye v. United States, 293 F. 1013 (D.C. Cir. 1923), which, although a criminal case, dismissed as evidence the polygraph machine because it had not gained "general acceptance" within its field, which at the time were the fields of physiology and psychology. This had been held as the standard for admissibility of theory and evidence for decades. But the Supreme Court under Daubert determined that "general acceptance" was not enough.

Take as an example a case in which a transformer or an electric motor is alleged to have been the origin of a fire that caused considerable damage. Plaintiff's expert witness may be an independent electrical engineer with a long list of credentials and previous experience as an expert witness in other, unrelated cases. The expert witness is commonly a college professor from the school of engineering.

During cross-examination, however, counsel for the defense could bring to light the fact that the engineer never specifically worked with, engineered, or worked for a company that built transformers or electric motors. Furthermore, it may also be disclosed that the expert never performed, much less documented, any failure-analysis test to prove out his theory. Therefore, the theory isn't considered scientific, nor is the expert any longer considered credible. The Supreme Court under Daubert heavily emphasized the role and responsibility trial court judges must play as "gatekeepers" in screening expert testimony for relevance and reliability, and not to automatically allow it to be presented to the jury simply because it is the opinion of a so-called "expert."

The Supreme Court explained that the expert's testimony must be "reliable"; that is, the opinion must have a "reliable basis in the knowledge and experience of the expert's discipline." The Daubert factors that the courts should employ in determining whether a theory or hypothesis constitutes "scientific knowledge" would include:

1. *Testing*: whether the expert's theory or technique can be, and has been, tested
2. *Peer review*: whether the theory or technique has been subject to peer review and publication
3. *Rate of error*: whether there is a known or potential rate of error
4. *Standards*: whether there exist, and have been maintained, standards controlling the technique's operation
5. *General acceptance*: whether the theory or technique is generally accepted within the relevant scientific community

Other significant rules and guidelines are the Federal Rules on Evidence 403, 702, 703 and 706.

Fed. R. Evid. 403: The expert testimony cannot have the tendency to confuse or mislead the jury. As applied in Daubert, "Expert evidence can be both powerful and quite misleading because of the difficulty in evaluating it. Because of the risk, the judge in weighing possible prejudice against probative force under Rule 403 of the present rules exercises more control over experts than over law witnesses."

Fed. R. Evid. 702: Scientific experts must possess the requisite knowledge and be sufficiently reliable in order to express his or her opinions to the jury.

Fed. R. Evid. 703: The facts and/or data relied upon by the expert must be "of a type reasonably relied upon by experts in the particular field in forming opinions or inferences upon the subject."

Fed. R. Evid. 706: The court also has the ability to chose its own expert in evaluating scientific testimony.

The bottom line in this analysis of the credibility of plaintiff's expert witness, and the fact that the testimony this witness is expected to produce doesn't comply with the preceding guidelines, allows counsel for the defense to ask for a Rule 104(a) hearing prior to admitting this testimony. This is a motion by the defense to have plaintiff's expert witness impeached by the judge. The defense tactic is to ask for a Rule 104(a) hearing late enough in the pretrial process to handicap the plaintiff, yet early enough to allow the court time to decide on the issue without leaving enough time for the plaintiff to obtain another expert.

Stanczyk v. Black & Decker, Inc., 836 F. Supp. 565, 566 (N.D. Ill. 1993): The court evaluated plaintiff's expert's testimony that he could design a guard for defendant's saw that would reduce possible contact with the exposed blade. Citing the expert's admission that, although he was confident that his design would work, his concept was not fully defined, fully proven, or fully documented. The court's analysis of this was a textbook example of the "gatekeeper" role prescribed by Daubert:

> Daubert teaches that the court must consider certain factors; the most important factor is whether the technique or theory being advanced by the expert can be or has been tested. The expert offered no testable design to support his concept. Furthermore, the history of engineering and science is filled with finely conceived ideas that are unworkable in practice. There is a high potential rate of error for mechanical concepts offered engineering analysis.
>
> One must consider whether there is peer review and publication of the technique. There is none, and the closest proxy for it, industry practice, produces no evidence of the use of gravity guards for this type of saw. This rules out the possibility of general acceptance, another factor to be considered. Finally, to the extent there are controlling design standards, they offer no support to the expert's opinion.

The court concluded that the expert's testimony did not pass the criteria under Daubert standards and excluded the evidence.

Pries v. Honda Motor Co. Ltd., 31 F.3d 543 (7th Cir. 1994): In an action arising from an automobile rollover, the district court granted summary judgment for the manufacturer. In that case, it was alleged that the occupant's seatbelt had come open during the course of the rollover accident. A test was performed by plaintiff's expert on a similar seatbelt latch by repeatedly dropping it on a hard surface to see whether it would open. Sometimes it did. Subsequently, when defense counsel asked what forces had brought about this opening and whether these were commonly achieved in a crash, the expert did not know. The court held that under Daubert, evidence of this kind is not scientific and does not satisfy Federal Rule of Evidence 702.

Fitzpatric v. Madonna (Pa. Sup. Ct. 1993): A 16-year-old boy was killed after being struck by the propeller of a motorboat. The jury awarded more than $1 million dollars in damages to the victim's mother in a wrongful death and survival action. The Pennsylvania Superior Court overturned the verdict. It found that the plaintiff's claim that a shroud over the propeller would have prevented the accident ignored the potential undesirable effects of such a device.

> An outboard motor is designed to move a boat through water. It has not been designed to allow motorboats to move among swimmers. The risk inherent in such movement is readily apparent. Moreover, it cannot be said with any degree of certainty that the risk of injury will be reduced by a safety guard, for the presence of a shroud over the propeller presents its own risks to swimmers. For example, a shroud creates a larger target area. In addition, the possibility exists that human limbs may become wedged between a shroud and the propeller, exposing a swimmer to even greater injury.
>
> A competent person knows that he or she must stay clear of the churning blades of an outboard motor in the

same way as a person avoids airplane propellers, chain saw teeth, and lawn mower blades.

If either party is successful in rebutting the other expert's analysis before or during trial, such a tactic would have substantial impact on the case or on that individual's credibility with the jury and the grounds for their case. This would be a devastating blow to the plaintiff's case. If counsel for the plaintiff is cognizant of such a potential move, they may try to change their facts and opinions just prior to court. If this should happen, and plaintiff's expert presents a whole different scenario, defense would want to move that the court either suppress this new theory or allow for additional depositions and discovery.

SELECTION OF TYPE OF TRIAL

When a product liability case does go to court, both parties are going to do their best to convince the court, specifically the jury, that they are right. Idealistically, both parties, or all the parties (in the case of multiple defendants), walk into court without any initial advantages or disadvantages, but this isn't always true. In some situations the defendant is looked at as an outsider by the members of the jury. Even worse, the plaintiff can be a well-known or locally liked personality. Or, if the plaintiff isn't personally known by the members of the jury, they could still look at the individual as one of their own, as opposed to the defendant, who could be viewed as a foreigner, especially since the defendant may well be from a different part of the country. In situations like this, prospects for a fair trial could be severely hampered.

And finally, maybe it's just a case of the poor little plaintiff against the corporate giants, the David versus Goliath syndrome. Counsel for the plaintiff will try to win the sympathy of the jury by making the plaintiff appear to be just like the rest of them, as well as an innocent victim of the defendant's negligence or possibly greed.

In all these scenarios the defense may start its case with a distinct handicap. One strategy for overcoming some of these disadvantages is to have the case moved from local state courts into the federal court system. Counsel for the plaintiff may not have a strong technical case, but will gamble on their chances with the jury composed of the plaintiff's local peers. But the defendant initiates a counterstrategy, and has the case moved from the local court into the federal court. This not only changes the complexion of the trial, but also increases the implied level of severity.

Moving a product liability case from a state court to a federal court has a distinct number of advantages, depending on the location. Although both courts will involve juries, the federal court will pull its jury members from a much larger district, which will help dissolve the potential for local prejudices. The new jury will comprise citizens not from a single county but from several counties. In the state courts, laws that govern what can be entered as evidence, statutes of limitations, and various other legal guidelines can be much looser than they will be in the federal court system, and not to the defendant's advantage.

There are certain other advantages to moving into the federal court. To begin with, the judges themselves tend to be more conservative then some of those in state courts. Although the jury is made up of regular citizens just as in the state system, it is felt that the jurors themselves take on a different attitude than those summoned for the local courts, that they take their position far more seriously because the fact that the case is in federal court indicates to them that it is a more serious event. With this in mind, they concentrate more on the case and the evidence presented, and less on any original prejudices.

On the other hand, there are also disadvantages to moving the case into the federal courthouse. Product liability cases are often rescheduled at the last minute, to make room for criminal trials. It wouldn't be uncommon to have a trial date nearing and, as you prepare and get psyched up, be notified by the court that your case was being pushed out another 3 months, because they brought in a federal drug case. And in 3 months it could happen again.

The decision to move the case into federal court should be the decision of defense counsel, especially if counsel was retained locally. The attorney should have the best feel as to whether the defendant will receive fair treatment in the local system with local judges or whether it is a better strategy to move the case. This is something that the insurance company and Liability Expert won't have any feel for, but the Liability Expert can discuss this move with counsel as part of their strategy planning.

Another prerogative would be to ask for a judge-only trial, as opposed to a jury trial. In a judge-only trial, both parties present their cases to a judge, who makes the final decision. A likely candidate for a judge-only trial would be a product liability case that was felt to be too technical for the average juror to comprehend. These kinds of trials can be held in the state civil court system as well as the federal court system. Some of the advantages of this type of trial are that the judge would be more likely to ask the right questions, demand the proper proof, and hopefully reach a just decision. This is not always a guarantee, however, for it is sometimes found that even a judge can be confused and overwhelmed by the technical nature of a case.

16

Case Studies

The intent of this chapter is first to demonstrate how the position and role of Product Liability Expert has been effective for some corporations in stopping potential product liability cases from materializing and second, to describe other product liability cases on record that newly assigned Liability Experts may learn from. There is much to be gained by studying the failures and successes of others rather than head into this field without such knowledge.

THE CLOCK FIRE

The first case study deals with a fire that broke out at about 3 a.m. in a small southern bar that had closed two hours earlier. The bar had a large front area and a newly added back room. A person driving by the bar noticed smoke coming out of it at around 5:30 a.m., and called the fire department. When they arrived, they found a fire burning along the backbar area, which had spread up onto the ceiling and was filling the bar with smoke and ash. Because the bar had brick outer walls, and the fire had not yet risen into the ceiling frame and roof, they were able to put it out rather quickly. The fire had still succeeded in damaging enough of the front area that the main bar could not be used until it was rebuilt. The owner was able to serve customers in the new back addition, however, once they passed through the fire-damaged front.

When the fire was out, the volunteer fire marshall inspected the bar to determine the source of the blaze. The bar consisted of one large room, about 30 feet square. Along the back wall was the bar, which was about 20 feet long. The backbar countertop and wall were the most heavily damaged sections, followed by the ceiling. On one end of the bar, about 8 feet from the center, was a bundle of electrical wires hanging down from the ceiling. Along the wall on the backbar were electrical circuit boxes and outlets. This area displayed some significant damage, along with 2 × 4 wall studs that seemed to have been the path for the fire to rise from the backbar surface up to the ceiling.

In the center of the bar was the cash register, and remnants of snack-bag displays and other paper products that were pretty well burned. To the left was a heavily blackened wall area, which showed every indication of an intense fire. This is where a decorative illuminated L.E.D. plastic clock had stood. Directly above this backbar area, the ceiling was burned, and the burn path led straight across the ceiling over to the electrical wires. In the brief investigation that followed, the fire marshall wrote on his report that the fire seemed to be the result of an electrical short. But that was it.

The owner of the bar contacted his insurance carrier to make them aware of what had just transpired. The insurance carrier sent one of their local representatives to the bar to inspect the damages and determine the financial loss, as well as to find out the cause. The insurance investigator read the fire report, which stated very little. Then the insurance investigator inspected the inside of the bar and the heavily burned backbar area. During this inspection, the investigator noticed the electrical wiring hanging down from the ceiling, and asked the owner what this led to. The owner explained that the wiring led to the new addition in back, and that a friend of his (who also happened to be present) was doing the wiring for him. The electrician friend was quick to interject that nothing he was doing could have caused this fire. The investigator then walked down the bar noticing all the paper materials along the bar top that had burned, and then quickly noticed the heavily blackened area with a radius of about 3 feet where the plastic clock had once stood, and inquired about what used to be there. The bar owner informed the investigator that a 10" × 14" plastic clock promoting a national product had been there, and the three decided that this looked like the probable source of the fire. The owner told the investigator that the clock had been brought in about 6 months ago by one of the bar's suppliers, who represented the national company advertised.

The next move was for the insurance company to retain the services of a local forensics expert, who went to the bar to verify probable cause. When the forensics investigator arrived, the bar owner and insurance investigator were also there. The forensics expert had brought along a video camera, and began videotaping the inside of the bar, narrating by describing what was being shown. He panned the whole inside of the bar and recorded the extent of the damage. Then he focused in on the right side of the backbar where the wires were hanging down from the ceiling, commented that this area was pretty heavily scarred, and asked what the dangling wires were there for. The insurance investigator briefly stated that they led to a new addition in back, but quickly added that none of this was the cause of the fire. He immediately led the videotaping forensics expert to the heavily scorched area where the clock had once stood and volunteered that this was where the fire started. The forensics expert commented on the intensity of the charred area, and inquired as to what had been there. The insurance rep said that it was a large illuminated electric clock, and the expert videotaped the whole area and agreed that this did look like the logical ignition point. After the inspection, the forensics expert furnished the insurance representative with a documented report of the findings, along with a copy of the video.

The insurance company's next move was to contact the national company advertised on the clock to make them aware of what their product had done. The legal department of the large national company replied that they had no direct responsibility, that they were indemnified against such incidents by the actual manufacturer of the clock, and that the insurance company needed to make the manufacturer aware of the liability. The manufacturer was a northern high-volume job shop that specialized in the manufacture of such clocks by the tens of thousands, along with several other types of promotional products. The insurance company sent the manufacturer a letter briefly explaining that one of their clocks had been found to be the cause of this specific fire, and that they were looking for subrogation in the amount of about $70,000. The letter asked that the check be sent to them as quickly as possible.

The manufacturer had their own Product Liability Expert in house, who was also the Quality Director and directed the efforts in their reliability testing laboratory. The Liability Expert reported the allegation to the manufacturer's risk insurance agent and noted that they were going to pursue an on-site investigation and meeting. With that, the Liability Expert contacted the bar's insurance carrier explained who he was and what his role is in the investigation of such incidents, and scheduled a meeting for the next week with the bar's insurance carrier, and the retained forensics expert. The risk carrier representing the manufacturer made arrangements to have one of their own local representatives present at this meeting as well.

The following week, the Product Liability Expert flew down to the location for the meeting, and met up with their risk carrier representative. Together they went into the meeting and met the bar insurance representative along with the forensics expert. The forensics expert first showed the fire report, which stated that the fire appeared to have started as a result of an electrical short (implying that the short was in this product, even though the report never mentioned the product). Then the forensics expert presented his own report, which specifically stated that the clock was the cause of the fire—that it apparently shorted out internally, which caused ignition to the plastic, and that the fire then spread up the electrical cord to the outlet, up the conduit to the ceiling, along the ceiling to the area where the electrical wires were hanging, and then back down to the papers and other items on the backbar. The clock could be identified as the source because it was the hottest spot of the fire and because of the classic V pattern it left charred on the wall.

The group then looked at the video, which showed the expert walking into the bar, focusing first on the wires hanging down from the ceiling and then, curiously enough, at a large number of open circuit boxes and panels that had had their circuit and outlet faces removed, which were against the backbar below the wires and heavily burned. And then on the tape the insurance rep can be seen leading the forensics expert away from that area and over to where the clock was, explaining that this was where the fire started. The forensics expert pans this area, and concludes that this is obviously the source of the fire. After the video, the expert's theory was discussed. Although the entire presentation was very convincing to all the other parties present, including the manufacturer's insurance rep, it did not convince the Liability Expert. In fact, several things were stated that didn't seem probable.

After this meeting, the group (with the exception of the bar's insurance rep, who had to be excused because of previous commitments) drove to the bar itself, which had since been repaired. The bar owner was available to explain where everything was and to answer questions. The bar owner confirmed for the Liability Expert that he had had the clock for about 6 months prior to the fire. The Liability Expert asked the bar owner if the clock (which had a built-in L.E.D. clock along with a 6-inch fluorescent lamp and illuminated display panel) had ever malfunctioned during this 6-month period or the night of the fire. This would include not keeping accurate time, the light malfunctioning, or any other unusual occurrence. The bar owner stated that the clock always ran perfectly. This was the first indication to the Liability Expert that the clock hadn't been defective. After the visit, the Liability Expert told his own insurance rep that he wanted to run some tests on other units, and that the insurance company shouldn't do anything until he got back in touch after the tests.

When the Liability Expert got back to his office, he had sample units brought into the reliability testing lab, and began subjecting them to every type of test that might cause them to short out or fail. The unit was already a UL-approved design, and all the electrical components were enclosed in sheetmetal. Furthermore, the plastic that came closest to any heat source was flame-retardant. The only thing that wasn't flame-retardant was the decorative outside housing, which was out of the reach of any possible ignition source. So the forensics expert's theory was improbable.

Many other aspects of the expert's theory didn't make much sense as well, such as the flame following the wiring, which led the Liability Expert to further question the other expert's credibility. In a sneak move, the Liability Expert called the other expert directly, said that he was impressed with his investigation, and might think about using him someday himself. The Liability Expert asked the other expert to send him his or curriculum vitae. The Expert also asked how much such an investigation would cost if the manufacturer ever sought his services, and the other expert told him around $300. The Liability Expert knew from this that very little effort had been put into this investigation; the bar's insurance company was merely trying to get a recognized "expert's" backing of their own theory. And when the expert's résumé arrived, the Liability Expert found that the individual was basically a building and bridge engineer with no electrical background.

After completing all the failure-mode testing, the Liability Expert contacted his insurance rep, apprised him of the findings, and stated that their position would be to fight the case. The Liability Expert concluded that the clock was nothing more than a casualty of the fire. The clock left such a charred impression on the wall purely because the outside plastic housing would burn much hotter than the other papers and materials on the backbar. But the fire didn't start with the clock; it may have started with the wiring and burned a path over to the clock. Or it may have started with something as simple as a cigarette left lying on the backbar. No one would ever know now that it had all been rebuilt. The insurance rep agreed, and sent a letter to the bar's insurance carrier advising them of all the testing and final results, and the fact that they would not accept responsibility and were fully prepared to go to court if necessary. The case died at that point. Had the Liability Expert not gotten involved, the manufacturer's insurance rep, who had already been convinced that the fault was theirs, probably would have settled.

THE NEON SIGN CASE

The company was a small manufacturer of custom neon window signs. They received notice that one of their signs allegedly shorted out and caused severe shock and injury to a young female employee of a convenience store. The letter alleged that the product was defective in nature and by design, and lacked adequate warning.The implication was that a court action was soon to follow unless the manufacturer showed signs that they were willing to compensate the party for her injuries and expenses. The letter did not specify a dollar amount. The manufacturer immediately notified their risk carrier of the notice and began to investigate the incident.

The manufacturer first sought to meet with the attorney representing the injured party, and to view the product. The manufacturer asked the carrier if they wanted to have anyone present for the meeting; they replied that the manufacturer should go ahead with the preliminary investigation on their own. The manufacturer's Liability Expert then contacted the attorney, and they arranged to meet the following week.

The Liability Expert flew down to the city where the attorney was, and they met at the convenience store where the accident took place. The employee who had been injured was receiving workers' compensation and would not be present for the meeting. The sign in question was a typical 2-foot-square neon window sign, with a skeleton metal frame that hung from ceiling hooks by two 3-foot chains, and an electrical cord that led back to the outlet. A neon sign is powered by a high-voltage transformer, normally 6000–9000-volt outputs, but with a regulated low 0.3 amps. If an individual receives a shock from a neon sign, the high voltage will deliver a good shock but it almost never results in serious injury. At the time, though, the stunned recipient thinks he's just almost been killed.

Neon signs are very basic in nature and design. The neon tube is attached to a metal frame, along with a heavy metal transformer. The transformer and sign are turned on and off by a small pull-chain exiting out the side of the transformer. Leads from the high-voltage transformer are connected to the ends of the neon tubing. The wire connections are covered by a rubber insulative sleeve, which can be slid out of position so that the wire can be untwisted and the tubing or the transformer replaced.

When the Liability Expert and the attorney met at the convenience store, the attorney had the neon sign with him. Because the glass tubing was broken, the sign would no longer light. The horizontal metal glass-support bar stock on one end was also bent downward. The attorney explained to the Expert that the employee was reaching for the pull-chain to turn the sign off when she suddenly received a severe shock, was at first unable to let go of the sign, and eventually broke free and was hurled back into a display case. Apparently unconscious, she was taken to the hospital, checked out by the doctors, and treated for the shock. The attorney also noted that the women had a burn mark on the inside of her left arm and on the sole of her left foot. These last observations were offered to demonstrate the severity of the electrical shock, but in fact it aroused the Expert's curiosity.

The attorney stated that the transformer obviously shorted out and that the whole metal structure was live, and when the employee tried to turn the sign off merely by pulling the chain, she was shocked. The Expert pointed out that if the transformer had a short, then it should still be defective. Even though the tubing was broke, it could still easily be determined if the sign or transformer was electrically shorted. This analysis was already beyond the comprehension of the attorney. The Expert plugged in the unit and checked whether the frame or transformer was live or shorted. Neither was found to be a problem, which eliminated the theory of a short. The Expert then tested to see if the transformer still worked, and found that it did. So now what the Expert had was a bent, broken, neon sign that still worked. In an inspection of the entire assembly, the Expert found that one of the insulative sleeves that covered the glass connection had somehow been slid out of place, and it was near the bend in the steel horizontal frame bar. This meant that the woman could have been shocked by the exposed connection. But how did it become exposed? And if she was only trying to turn it off, how would she have come into contact with the electrical connection? And how did the glass get broken? The attorney's explanation was that the sign obviously fell to the floor during all the excitement, breaking the glass and bending the frame. The Expert wasn't satisfied; things didn't add up yet.

The Expert reflected back on the doctor's findings and asked, "How could the woman have received a burn mark on the bottom of her left foot, which implies an entry or exit wound from electricity, when she was probably standing on the floor with her shoes on?" The attorney didn't have an answer, so he called over the store owner and asked what specifically the woman was doing when she supposedly was turning off the sign. The store owner said the employee had been sent to the front of the store to wash the windows. The Expert then asked how she would be able to reach all of such a large window. The store owner said that the employee was standing on a stepladder to wash the window when she received the shock. The original story was starting to unravel.

The Expert asked, "If she was standing on a wooden ladder, how would she have received an electrical shock since she wasn't grounded?" The store owner said it was actually an aluminum ladder. "What kind of shoes was the woman wearing?" asked the Expert. "She wasn't wearing any shoes," replied the owner; it was a hot day and she was barefoot. Things were beginning to come together for the Expert. He now had a barefoot woman, standing on a metal ladder, washing windows with wet hands and sponges or rags. What didn't make any sense yet was why the woman would at the same time have been turning off the neon sign. And how the metal frame would have been bent, the glass broken, and burns sustained on the bottom of her left foot and inside her left arm.

The Expert thought about the attorney's story, that when the woman became shocked she couldn't let go of the sign and was thrown backward into a display case. The

Expert knew from experience that people who are shocked by these high-voltage, low-amp transformers immediately jump free from them. They are not immobilized and unable to break loose. When the Expert asked where the sign had been hung, the owner pointed to a spot about 6 feet off the floor. Two hooks were still screwed into the ceiling. The chains, the Expert knew, hung down from these hooks and were fastened to the metal frame of the sign by S-hooks. This suspension system was strong enough to hold a person, so it was unlikely that the sign broke free and dropped to the floor. There must be some other explanation for the broken glass and the bent frame.

The Expert put together a new scenario for the attorney: “the employee was standing barefoot on the metal stepladder washing the front store window, but she didn’t try to turn the sign off. Instead, she was reaching way out to her left while on the ladder, so far that she was standing only on her left foot. Then she apparently lost her balance, quickly grabbed on to the left side of the sign frame to catch her fall, which caused the right side of the sign to snap up under her left arm, thereby disturbing the insulative sleeve and exposing the connection to the inside of her left arm. This accounts for the electrical entrance burns on the inside of the left arm and at the bottom of the left foot. The reason she couldn’t let go of the sign wasn’t because of the electrical current but because she was off balance and would fall to the floor, which she eventually did anyway. And that is how this whole thing happened. It was merely an accident, or negligence on the part of your client, but not a defective product.” The attorney, somewhat amazed by the detective work, said merely that he would have to go back and talk to his client. The Expert never heard from the attorney again, and no one from the insurance carrier ever had to get involved.

It was a pretty impressive bit of investigation, but it couldn’t have been put together without a thorough knowledge of the sign and its mechanical parts. Had another attorney gotten involved, or had the insurance company handled it merely by sending an investigator, the focus of the investigation would probably have been the extent of the injuries, not the product and the scenario that led up to the injury. But that was where the real story was.

These two examples demonstrate the effective role played by the manufacturers' own Product Liability Experts. Their technical product knowledge combined with a thorough analysis of the actual events led to the conclusions that put an end to the action—that is, stopped it in its tracks, which is the objective of this new function.

THE MENUBOARD FIRE

The manufacturer of wall-mounted illuminated menuboards for fast-food restaurants receives notice from an insurance carrier representing a popular restaurant chain that their menuboard has been identified as the source of a fire that completely destroyed a $1 million-plus restaurant. This has been determined by the fire inspector as well as an independent insurance investigation company and a forensics expert hired by the restaurant's insurance carrier. In all, three separate investigations implicated the menuboard as the source and cause.

The menuboard is roughly 8 feet long and 2 feet high. It is internally lit by a number of fluorescent lamps powered by a large fluorescent-lamp ballast. The front of the menuboard is loaded with grids containing plastic menu strips and transparencies. The rest of the light box is composed of sheetmetal and aluminum framework.

The employees had arrived at the restaurant at around 5:00 a.m. By 6:00 a.m. they began to smell smoke but couldn't tell where it was coming from. As the minutes ticked by, the smell of smoke began to get stronger, so the employees called the fire department. After the firefighters had arrived, one opened an overhead trapdoor leading into an unused attic, and almost instantaneously the roof was engulfed in flames. By the time the firefighters were able to put it out, the restaurant was a total loss.

The menuboard had been mounted to a wooden wall above a window through which food was passed from the stove area to the front counter. Below the menuboard and at the top of the window frame were heat lamps to keep standing food warm. By the time the fire was extinguished, the front of the menuboard had been destroyed and the wall was heavily charred in a V pattern, which seemed to start in the bottom center of the menuboard (where the fluorescent-lamp ballast was mounted) and went straight up the inside of the wall into the ceiling and beyond. The reports by all three independent investigators have indicated that the fire started with the ballast—apparently as a result of its shorting out, igniting various plastic strips and transparencies on the menuboard front—and traveled up into the ceiling and attic space, until it was finally out of control.

By the time the manufacturer was notified, the restaurant had been leveled and a new one built. A unique thing that had happened, however, in preparation for possible litigation, was that the restaurant's insurance carrier had had what was left of the menuboard and the wall removed, crated up, and stored at a remote warehouse.

When the manufacturer received notice, their Director of Product Reliability, who functioned as the Product Liability Expert, immediately had their insurance representatives contact the restaurant's insurance carrier and make arrangements for a forensics investigation. In addition to the manufacturer's own Expert, the manufacturer's insurance carrier retained the services of a local forensics expert and the restaurant hired another forensics expert to represent them. A meeting at the warehouse was then set up, with the three experts to attend. Once everyone was there, the warehouse door was opened by an attendant, and there was the sealed crate containing the remainder of the wood wall and menuboard. The crating wood was removed, exposing the charred contents. Little was left of the aluminum menuboard. The sheetmetal back was still screwed to the plywood-paneled wall, although blackened and covered with rust. The menugrids were completely gone, but they had been removable to begin with. The lamps had been broken and were gone. The remainder of the menuboard and its back were unfastened from the wall and laid out on the ground. The plywood face of the wall displayed the heavy V-shaped burn from the very bottom of the menuboard center, actually lower than where the menuboard was, straight up to what must have been the ceiling.

The experts all agreed that the wall showed the classic burn pattern for a point of origination. Each of the experts was also loaded with camera equipment, and each time a screw was turned or an object moved, scores of pictures were taken. They next studied the remains of the electrical wiring in the lightbox. Although the front of the menuboard was no longer intact, the aluminum frame was still there, as was the wiring lying in the bottom of the board running from the ballast to the fluorescent lampholders, although everything was heavily burned. But that is probably the best thing about electrical wiring: no matter how burned everything else is, the wires and wire connectors are normally still present.

The experts traced the wires running along the bottom of the menuboard to the individual lampholders. Although the insulation was totally burned away, each of the experts knew that, had the wires shorted anywhere, there would still be a whitened arc spot. After tracing all the wiring, they noticed no such point of arcing. The last item that needed to be analyzed was the ballast. Although the two outside experts were quite knowledgeable about electrical components and how ballasts are made, the Director of Product Reliability had more specific product knowledge; he knew exactly what a ballast would look like if it had been arcing or radically shorting out in a manner that would have led to a fire. In either scenario, the ballast would have had either a hole blown through its side or a whitened area along its seams. Even though the ballast in question was blackened by the fire, neither condition was present. The Product Reliability Expert almost instantly knew that the ballast was not at fault for the fire, which meant that neither was the menuboard.

The two forensics experts meanwhile removed the ballast from the aluminum frame and packed it into a box to be analyzed at the nearest lab. At the lab, they began the 4-hour process of tearing down the ballast, photographing every step. In the end, all three experts agreed that the internal workings of the high-voltage ballast showed no abnormal signs of shorting out, and therefore the menuboard was eliminated as the source of the fire.

Although there was no way to prove it with the evidence destroyed, the source was probably the heat lamps mounted to the bottom of the wall just below the menuboard. One good indication of this was that the V-pattern fire actually went a few inches lower than the menuboard, the location of the tops of the heat lamps. With all the unusual efforts to crate up most of the wall, the heat lamps hadn't been secured, and so the actual cause would never be confirmed.

The forensics experts might have come to the same conclusion whether or not the Product Reliability Expert had been there, but with his representing the manufacturer and instantly indicating that the ballast wasn't defective, he may also have been the driving force that led to the intricate breakdown of the unit. It should also be noted that one of the forensics experts had originally been hired by the restaurant's insurance carrier to conduct an investigation at the fire scene, and reported that the ballast was the cause of the fire. Now that same individual would have to go back to his client and recant. In the end, the manufacturer never heard about the incident again and closed their files on it.

From a less in-depth perspective, it is also educational for companies to learn about other product liability cases, to appreciate what they need to take a second look at and where to focus their efforts in prevention. The following are brief summaries of lawsuits that were initiated, which means, regardless of how they ended up, the company and its insurance carrier had to get involved, hire attorneys, and begin to defend themselves.

THE KITCHEN RANGE CASE

The company being prosecuted is a manufacturer of kitchen ranges, including the range hood. Two young brothers were fighting when one suddenly pushed the other toward the stove, causing him to hit his head on the sharp corner of the range hood. His head was cut open, which resulted in a serious injury. The parents took their son to the hospital to be treated, and eventually hired a lawyer to represent them in suing the manufacturer of the range. Their complaint alleged defective design and reasonably foreseeable risk of harm. Although the defense would surely be able to claim contributory negligence, the manufacturer might still be found partially negligent themselves because if the range hood had been designed with rounded corners instead of sharp pointed ones, the injury would not have occurred. And even though in this case the injury was attributed to horseplay, counsel for the plaintiff could argue that the defendant could just as easily have tripped and fallen into the range hood and suffered the same injury—thus, the issue is design and not the actions that led to the injury, all of which could fall under the category of foreseeability.

THE WATER HEATER CASE

In this case, the defendant is the manufacturer of a gas hot-water heater. A 16-year-old boy, having finished painting the house, went into the utility room to clean the brushes and trays. While he was washing out the brushes with gasoline, the hot-water heater kicked in and the fumes exploded into flames. The boy received serious burns all over his body. The water heater manufacturer is sued for lack of warning—

more specifically, failure to warn that the unit should not be placed in an area where gasoline or flammable liquids or fumes may be present.

THE HAIR CURL RELAXER

A woman applied a commercial hair curl relaxer to her hair and went to bed. The next morning, when she lit a cigarette, her head and hair burst into flames and she suffered burns to her scalp and head. She sued the hair relaxer company for $450,000 for not having any warnings on the bottle that the relaxer is flammable.

THE UNBELIEVABLE LAUNDROMAT CASE

The defendant in this case is a manufacturer of laundromat driers. The scene takes place in January in downtown Chicago. The temperature had plummeted well below 0°. Some homeless men were looking for a warm place to sleep. The laundromat is open 24 hours a day, so late at night some of the men moved into the place, climbed inside the dryers to stay warm, and went to sleep. The next morning, some local kids on their way to school cut through the laundromat to save time. As they passed by the dryers, they popped quarters into the slots, pushed the start buttons, and left. The homeless men were left tumbling in the dryers.

One of the tumble-dried men decided to visit a local lawyer and initiate a lawsuit against the dryer manufacturer, citing lack of warning that a person could be injured if he climbed inside the dryer.

OTHER CASES OF RECORD

Sears Roebuck & Co. v. Harris, Ala SupCt, No. 1911519: Plaintiff installed a used water heater in his grandmother's mobile home, but failed to vent the heater properly before connecting it to a liquid propane gas supply. Liquid propane produces large amounts of carbon dioxide due to incomplete combustion. Four members of the family spent the night in the mobile home and died from carbon monoxide poisoning. The plaintiffs sued the heater designer, the manufacturer, and the store chain that sold the unit, alleging that the heater was unreasonably dangerous because it was not equipped with a carbon monoxide sensor or vent safety shut-off device that would have prevented the poisonings. The claims were brought under the Alabama Extended Manufacturer's Liability Doctrine, and on the theory of negligent or wanton failure to warn.

The court ruled that the improper installation was a reasonably foreseeable misuse, and further evidence indicated that the defendants were knowledgeable about venting failures that resulted in serious injuries. The courts further stated that the defendants had a duty to anticipate reasonably foreseeable ways consumers would misinstall their product, which was further compounded by the fact that the water heater had a label promising "You can install it yourself," even though the manufacturers knew that customers had in fact installed them improperly and yet failed to make changes to the label. The court awarded the plaintiffs $6.5 million.

Gillham v. Admiral Corp., 523 F.2d 102 (6th Cir. 1975): Punitive damages were awarded against a television manufacturer when a set caught on fire. It was proven in court that the manufacturer had prior knowledge that its transformers had the potential to catch fire because of defective design, even before the television sets were taken to market. It was also found that shortly after the sets entered the marketplace, and years before the plaintiff suffered injuries, the manufacturer had been informed of the problem. Despite this informed knowledge, and knowing that some of the sets were indeed catching fire even when not in use, the manufacturer made no effort to redesign the set or inform purchasers about the hazard. Instead, the manufacturer continued to market the set and make assurances that it was safe.

Gellenbeck v. Sears Roebuck & Co., 59 Mich. App. 339, 229 N.W.2d 443 (1975): A child was injured when a swing chain broke while she was being twirled on the swing by other children. The chain was found to have strong load-bearing capacity but low resistance to twisting. Because it was foreseeable that the chains would be used in swing sets, liability against the manufacturer on a warranty theory was upheld.

Darson v. Globe Slicing Machine Co, 200 A.D.2d 551, 606 N.Y.S.2d 317 (2d Dept. 1994): A 14-year-old plaintiff sued the manufacturer of a meat grinding machine because of injuries sustained while assisting in the operation of the machine. A trial court set aside a jury verdict for the plaintiff and dismissed the case. The case went to appeal. The appellate division re-affirmed and held that, pursuant to Labor Law 133(1)(c), it is illegal for persons under the age of 16 to assist in the operation of such machinery, and that the plaintiff was therefore not a reasonably foreseeable user. The appellate court further noted that the manufacturer could not have foreseen that the purchaser would discard the warnings that came with the piece of equipment without passing those warnings on to its employees.

Meyerhoff v. Michelin Tire Corp., 852 F. Supp. 933 (D. Kan. 1994): A truck driver (plaintiff's decedent) was killed when a Michelin truck tire he was attempting to repair, reinflate, and remount exploded. His parents brought an action against the tire company, alleging that they failed to adequately warn their son of the dangers involved in working on the tire. Plaintiffs argued that the following warning should have been placed on the side of the tire in yellow or some other contrasting color:

> WARNING: TIRE MAY EXPLODE WHEN REINFLATED, CAUSING SERIOUS INJURY OR DEATH. DO NOT REINFLATE AFTER RUNNING UNDERINFLATED. TAKE THE TIRE TO A MICHELIN DEALER FOR REPAIR.

The jury found that Michelin was at fault for failing to place a warning on its tire. In posttrial motions, Michelin contended that "plaintiffs produced no evidence that it could have placed an adequate warning on the sidewall of the truck tire that exploded and killed Kevin Meyerhoff or that its failure to place a warning on the sidewall of the truck tire caused or contributed to Kevin Meyerhoff's injury."

At the trial, the plaintiffs called an expert to testify about the feasibility of placing warnings on a truck tire's sidewall. The expert, however, was unable to testify that it would have been feasible for the warning to be in yellow or another contrasting color, particularly a warning with lettering of the size proposed by plaintiffs. Accordingly, the jury was instructed that, while it could consider the wording of the proposed warning, it could not consider any evidence concerning the color.

Plaintiffs next called the expert who authored the proposed warning to testify about the adequacy of truck tires' sidewall warnings. He testified that the best way to convey a warning is to "put it on the product itself." He also testified about warnings other manufacturers placed on sidewalls, which he declared to be inadequate but "definitely better than nothing." Indeed, on cross-examination, plaintiff's expert admitted that, in his opinion, every warning he had observed on other manufacturers' tires, whether for trucks, passenger cars, or farm tractors, was inadequate. Thus, "there were no adequate tire sidewall warnings in use at the time the Michelin truck tire was sold." He also admitted that there were many dangers associated with truck tires that the language of his own proposed warning failed to mention.

Plaintiff's expert was then asked during cross-examination about an experiment he conducted to evaluate the effectiveness of his proposed "yellow warning." He had fabricated a tire with his warning and taken it to three repair shops, where he presented the tire for inflation and asked the repairmen about the meaning of the warning. None of the three repairmen questioned was able to answer simple questions about the meaning of the yellow warning on the tire. The expert terminated the experiment because it was not productive.

Kansas law, however, has a "read and heed" presumption. The jury was instructed that "there is a presumption that an adequate warning will be read and heeded," and also that there is a "presumption that an inadequate warning caused the injuries." The district court felt itself bound by Kansas law on that subject, and therefore held that it could do nothing about the jury's finding on causation. The court did, however, do something about the jury's finding on the underlying claim that Michelin could and should have placed an adequate warning on the tire. Applying a Daubert analysis, the court found there was no legimate foundation for the expert's testimony on failure to warn, and therefore no reasonable basis for the jury's finding of fault. Specifically, the court found that plaintiff's expert had never shown that their proposed "adequate" warning could in fact be implemented and that it would be effective.

> In this court's view, a person cannot, after suffering an accident, simply draw up a warning limited to the dangers involved in that accident and argue that that warning warning should have been conveyed by the manufacturer or seller without first also establishing that that warning is adequate and that it actually could have been communicated in the manner proposed.
>
> In a warning case, a plaintiff must do more than simply present an expert who espouses a new or different warning. He must establish that warning's feasibility, adequacy, and effectiveness. In this case, plaintiff's experts' testimony fell woefully short of meeting these criteria.

The court concluded that the testimony of plaintiff's experts was "largely unsupported speculation" that "lacked the underpinnings of scientific validity demanded by Daubert." As a result, the court granted Michelin's motion for judgment as a matter of law.

Fitzgerald v. Caterpillar Tractor Co., 683 S.W.2d 162 (Tex. Ct. App. 1985): The plaintiff was injured when the blade of a forklift became disengaged and fell on his foot. Plaintiff sued the manufacturer alleging breach of implied warranties of fitness and merchantability. The trial court granted a directed verdict in favor of the defendant. On appeal, this action was affirmed, since the plaintiff failed to meet his burden of showing that "the product in question was not merchantable or fit for the purpose for which it was intended on the date of delivery, or in other words at time of the manufacture of the product." The only testimony supporting the claim that the product was defective was from the injured plaintiff himself, stating that the forklift would be safer if it incorporated a certain safety feature. The appeal court agreed that the trial judge was correct in discounting that testimony and granting the directed verdict.

Sturm v. Clark Equipment Co., 547 F. Sup. 144 (W.D. Mo. 1982): In this case, the court permitted evidence that there had been only one accident in 74,000 machine-years of usage of 34,000 machines to demonstrate the reasonableness of the design and to rebut allegations of design defect. The defense ultimately won the verdict. If defense counsel can prove that the accident at issue was an exceedingly rare occurrence, the argument should be made that the "proof is in the pudding"; the adequacy and propriety of the design speak for themselves.

Myers v. American Seating Co., 637 So.2d 771 (la.App. 1st Cir. 1994): A school employee used a folding chair as a stepstool to reach a high shelf. The folding chair "jackknifed" when the woman moved toward the rear of the seat, i.e., aft of the pivot point. Noting that the manufacturer is not responsible to account for every "conceivable foreseeable use," the court found that standing on the rear of the seat was not a "reasonably anticipated use," even though it may have been a conceivable use: "Most people who use a folding chair as a stepladder utilize the front portion of the seat upon which to stand." The court considered the frequency of established incidents involving the seat, noting that only two incidents had occurred with more than 5 million chairs. Based on this, the court ruled as a matter of law that the use was not reasonably foreseeable, and the manufacturer was not liable.

Smits v. E-Z Por Corp., 365 N.W.2d 352 (Minn. App. 1985): Plaintiff was burned with hot grease when a light-weight disposable aluminum roasting pan buckled as she was removing it, and the roasted turkey it contained, from an oven. The pan was marketed with a label stating that it would hold up to 15 pounds. Also on the label, however, was a detailed instruction with cautionary language about supporting the pan's bottom when removing it from the oven. The court rejected the plaintiff's suggestion that she was entitled to a directed verdict on the express warranty claim, and affirmed the entry of judgment for defendant as entered on a jury verdict.

Montgomery Elevator Co. v. McCullough, 676 S.W.2d 776 (Ky. 1984): A 10-year-old boy sustained a crushed toe when his shoe caught between the treads and side skirt of an escalator. The manufacturer had sent modification kits to owners of that model of escalator, but the owner's failure to use the kit does not insulate the manufacturer from liability. Under 402A of the Restatement (Second) of Torts, judgment for the plaintiff was affirmed.

Caterpillar Tractor Co. v. Ford, 406 So. 2d 854 (Ala. 1981): A tractor driver was killed when the vehicle rolled over and crushed him. His widow brought suit against the manufacturer under Alabama's Extended Manufacturer's Liability Doctrine. The court held that there was adequate evidence to support a jury finding that the product was defective, since it was supplied without a protective structure to shield the driver in the event of a rollover. The availability of that kind of protection as an optional equipment item was not an adequate defense for the manufacturer.

Reese v. Mercury Marine Division of Brunswick Corp, 793 F.2d 1416 (5th Cir. 1986): A wrongful-death action was brought against the manufacturer of an outboard motor, alleging that the plaintiff's son would not have died in a boating accident if the motor had been equipped with a "kill switch" to cut off power if the boat operator was thrown or ejected from the operating position. Judgment was for the plaintiff, in an award of over $2 million.

Baker v. Promark Products West, Inc., 692 S.W.2d 844 (Tenn. 1985): The plaintiff was injured by a stump grinder, and sought to impose liability on its lessor and distributor based on theories of strict liability for defective design and inadequate warnings, and breach of the implied warranty of merchantability. The court held that the manufacturer was the only defendant properly exposed to a cause of action in strict liability, but that if it was shown that the manufacturer was insolvent, then the lessor and distributor could be subjected to liability on the basis of strict liability.

Briney v. Sears Roebuck & Co., 782 F.2d 585 (6th Cir. 1986): The plaintiff sustained serious injuries when one of his hands was pulled into the blade of a table saw while he was attempting to construct a combination mailbox and planter. The district court reasoned that, because the plaintiff had removed a blade guard, he had substantially changed the product and thus could not claim that the product had hurt him due to a defect. An appellate court rejected this reasoning, because the plaintiff's theory of defect was that the design of the product made it likely that a consumer would remove the blade guard since the guard prevented the use of the machine for certain kinds of cuts that it was intended to perform. Therefore, the problematic interference between the saw blade and the blade guard was not the result of substantial change but was "inherent in the manufacturer's design."

Liebeck v. McDonald's (New Mexico 1994): Plaintiff was scalded by coffee, causing third-degree burns. Liebeck was a passenger in a car driven by her grandson. At a McDonald's drive-up window in Albuquerque, plaintiff accidently spilled the coffee into her lap, resulting in the burns. Lawyers for the plaintiff contended that the coffee was too hot, 165–170°;

coffee brewed at home is generally 135–140°. Defendants expressed no willingness during the trial to turn down the heat or print any warnings. Plaintiff was awarded $2.9 million. Case was later appealed, and the award was lowered to around $0.5 million.

Glossary

A number of the legal terms that are commonly used and referenced in the field of product liability need to be defined for the Product Liability Expert, as well as for other members of the corporate team. Many of these terms can be found within this book, but there are others that, although not used here, should be known. It will also be noticed that some terms were defined by the courts in various legal cases cited. Dictionaries of law have the difficult job of defining these ever-changing terms, because the legal system itself is forever changing. Most of the definitions in this glossary are taken with permission from *Black's Law Dictionary*.

Affidavit A written or printed declaration or statement of facts, made voluntarily, and confirmed by the oath or affirmation of the party making it, taken before a person having authority to administer such oath or affirmation. (State v. Knight, 219 Kan. 863, 549 P.2d 1397, 1401.)

Attorney–client privilege In law of evidence, client's privilege to refuse to disclose and to prevent any other person from disclosing confidential communications between him and his attorney. Such privilege protects communications between him and his attorney. Such privilege protects communications between attorney and client made for the purpose of furnishing or obtaining professional legal advice or assistance. (Levingston v. Allis-Chalmers Corp., D.C.Miss., 109 F.R.D. 546, 550.) That privilege which permits an attorney to refuse to testify as to communications from client to him though it belongs to client, not to attorney, and hence client may waive it. In federal courts, state law is applied with respect to such privilege. (Fed. Evid. Rule 501.)

Breach of warranty In real property law and the law of insurance, the failure or falsehood of an affirmative promise or statement, or the nonperformance of an executory stipulation. As used in the law of sales, breach of warranty, unlike fraud, does not involve guilty knowledge, and rests on contract. Under Uniform Commercial Code consists of a violation of either an express or implied warranty relating to title, quality, content or condition of goods sold for which an action in contract will lie. (U.C.C. 2-312 et seq.)

Case law The aggregate of reported cases as forming a body of jurisprudence, or the law of a particular subject as evidenced or formed by the adjudged cases, in distinction to statutes and other sources of law. It includes the aggregate of reported cases that interpret statutes, regulations, and constitutional provisions.

Civil action Action brought to enforce, redress, or protect private rights. In general, all types of legal actions other than criminal proceedings. (Gilliken v. Gilliken, 248 N.C. 710, 104 S.E.2d 861, 863.) In the great majority of states which have adopted rules or codes of civil procedure as patterned on the Federal Rules of Civil Procedure, there is only of action known as a "civil action." The former distinctions between actions at law and suits in equity, and the separate forms of those actions and suits, have been abolished. (Rule of Civil Proc. 2; New York CPLR 103(a).)

Complaint The original or initial pleading by which an action is commenced under codes or Rules of Civil Procedure. (Fed. R. Civil P. 3.) The pleading which sets forth a claim for relief. Such claimant shall contain: (1) a short and plain statement of the grounds upon which the court's jurisdiction depends, unless the court already has jurisdiction and the claim needs no new grounds of jurisdiction to support it, (2) a short and plain statement of the claim showing that the pleader is entitled to relief, and (3) a demand for judgment for the relief to which he deems himself entitled. Relief in the alternative or of several different types may be demanded. (Fed. R. Civil P. 8(a).)

Contributory negligence The act or omission amounting to want of ordinary care on part of complaining party, which, concurring with defendant's negligence, is proximate cause of injury. (Honacker v. Crutchfield, 247 KY. 495, 57 S.W.2d 502.) Conduct by a plaintiff which is below the standard to which he is legally required to conform for his own protection and which is a contributing cause which cooperates with the negligence of the defendant in causing the plaintiff's harm. (Li v. Yellow Cab Co. of California, 13 Cal. 3d 804, 119 Cal. Rptr. 858, 864, 532 P.2d 1226.)

Curriculum vitae A résumé or summary of one's curricular background and experience. When used to qualify an expert witness, would include the expert's educational background, experience in the field, job history, and professional affiliations as they relate to the subject at hand.

Defendant The person defending or denying; the party against whom relief or recovery is sought in an action or suit or the accused in a criminal case.

Deposition The testimony of a witness taken upon oral question or written interrogatories, not in open court, put in pursuance of a commission to take testimony issue by a court, or under a general law or court rule on the subject, and reduced to writing and duly authenticated, and intended to be used in preparation and upon the trial of a civil action or criminal prosecution. A pretrial discovery device by which one party (through his or her attorney) asks oral questions of the other party or of a witness for the other party. The person who is deposed is called the deponent. The deposition is conducted under oath outside of the courtroom, usually in one of the lawyer's offices. A transcript—word-for-word account—is made of the deposition. Testimony of the witness, taken in writing, under oath or affirmation, before some judicial officer in answer to questions or interrogatories. (Fed. R. Civil P. 26 et seq.; Fed. R. Crim. P. 15.)

Disclaimer The repudiation or renunciation of a claim or power vested in a person or which he had formerly alleged to be his. The refusal or rejection of an estate or right offered to a person. The disavowal, denial, or renunciation of an interest, right, or property imputed to a person or alleged to be his. "Disclaimer of warranties" is a means of controlling liability of seller by reducing number of situations in which seller can be in breach of contract terms. (Collins Radio Co., Del. Super., 515 A.2d 163, 171.)

Discovery The pretrial devices that can be used by one party to obtain facts and information about the case from the other party in order to assist the party's preparation for trial. Under Federal Rules of Civil Procedure (and in states which have adopted rules patterned on such), tools of discovery include: depositions upon oral and written questions, written interrogatories, production of documents or things, permission to enter upon land or other property, physical and mental examinations and requests for admission. (Rules 26–37.) Term generally refers to disclosure by defendant of facts, deeds, documents or other things which are in his exclusive knowledge or possession and which are necessary to party seeking discovery as a part of a cause of action pending, or to be brought in another court, or as evidence of his rights or title in such proceeding. (Hardenbergh v. Both, 247 Iowa 153, 73 N.W.2d 103, 106.)

Et al. Meaning "and others." Often affixed to the name of the person first mentioned where there are several plaintiffs, grantors, persons addressed, etc. Where the words are used in a judgment against defendants, the quoted words include all defendants. (Williams v. Williams, 25 Tenn.App. 290, 156 S.W.2d 363, 369.)

Expert witness One who by reason of education or specialized experience possesses superior knowledge respecting a subject about which persons having no particular training are incapable of forming an accurate opinion or deducting correct conclusions. Kim Mfg., Inc. v. Superior Metal Treating, Inc., Mo. App., 537 S.W.2d 424, 428. A witness who has been qualified as an expert and who thereby will be allowed to assist the jury in understanding complicated and technical subjects not within the understanding of the average layperson. One skilled in any particular art, trade, or profession, being possessed of peculiar knowledge concerning the same, and one who has given subject in question particular study, practice, or observation.

Express warranty A promise, ancillary to an underlying sales agreement, which is included in the written or oral terms of the sales agreement under which the promisor assures the quality, description, or performance of the goods. It is not necessary to the creation of an express warranty that the seller use formal words such as "warrant" or "guarantee" or that he have a specific intention to make a warranty, but an affirmation merely of the value of the goods or a statement purporting to be merely the seller's opinion or commendation of the goods does not create a warranty. (U.C.C. 2-313.)

Full warranty A warranty as to full performance covering generally both labor and materials. Under a full warranty, the warrantor must remedy the consumer product within a reasonable time and without charge after notice of a defect or malfunction. (15 U.S.C.A. 2304.)

Implied warranty A promise or contract not written or stated. Exists when the law derives it by implication or inference from the nature of the transaction or the relative situation or circumstances of the parties. (Great Atlantic & Pacific Tea Co. v. Walker, Tex. Civ. App., 104 S.W.2d 627, 632.)

Indemnification In corporate law, the practice by which corporations pay expenses of officers or directors who are named as defendants in litigation relating to corporate affairs. In some instances corporations may indemnify officers and directors for fines, judgments, or amounts paid in settlement as well as expenses.

Indemnify To restore the victim of a loss, in whole or in part, by payment, repair, or replacement. To hold harmless; to secure against loss or damage; to give security for the reimbursement of a person in case of an anticipated loss falling upon him.

Interrogatories A pretrial discovery device consisting of written questions about the case submitted by one party to the other party or witness. The answers to the interrogatories are usually given under oath; i.e., the person answering the questions signs a sworn statement that the answers are true. (Fed. R. Civil P. 33.)

Litigation A lawsuit. Legal action, including all proceedings therein. Contest in a court of law for the purpose of enforcing a right or seeking a remedy. A judicial contest, a judicial controversy, a suit at law.

Negligence The failure to use such care as a reasonably prudent and careful person would use under similar circumstances; it is the doing of some act which a person of ordinary prudence would not have done under similar circumstances or failure to do what a person of ordinary prudence would have done under similar circumstances. (Amoco Chemical Corp. v. Hill, Del. Super., 318 A.2d 614, 617.)

Plaintiff A person who brings an action; the party who complains or sues in a civil action and is so named on the record. A person who seeks remedial relief for an injury to rights; it designates a complainant. (City of Vancouver v. Jarvis, 76 Wash. 2d 110, 455 P.2d 591, 593.)

Privity of contract That connection or relationship which exists between two or more contracting parties. It was traditionally essential to the maintenance of an action on any contract that there should subsist such privity between the plaintiff and defendant in respect of the matter sued on. However, the absence of privity as a defense in actions for damages in contract and tort actions is generally no longer viable with the enactment of warranty statutes, acceptance by states of doctrine of strict liability, and court decisions (e.g., MacPherson v. Buick Motor Co. 217 N.Y. 382, 111 N.E. 1050) which have extended the right to sue for injuries or damages to third-party beneficiaries, and even innocent bystanders (Elmore v. American Motors Corp., 70 Cal. 2d 578, 75 Cal. Rptr.652, 451 P.2d 84).

State of the art In context of product liability case, means level of pertinent scientific and technical knowledge existing at time of manufacture. (Wiska v. St. Stanislaus Social Club, Inc., 7 Mass. App. 813, 390 N.E.2d 1133, 1138, 3 A.L.R.4th 480.)

Statute of limitations Statutes of the federal goverment and various states setting maximum time periods during which certain actions can be brought or rights enforced. After the time period set out in the applicable statute of limitations has run, no legal action can be brought regardless of whether any cause of action ever existed.

Statute of repose "Statutes of limitations" extinguish, after period of time, right to prosecute accrued cause of action; "statute of repose," by contrast, limits potential liability by limiting time during which cause of action can arise. (Kline v. J.I. Case Co., D.C.Ill., 520 F.Supp. 564, 567.)

Strict liability A concept applied by the courts in product liability cases in which seller is liable for any and all defective or hazardous products which unduly threaten a consumer's personal safety. This doctrine poses strict liability on one who sells product in defective condition unreasonably dangerous to user or consumer for harm caused to ultimate user or consumer if seller is engaged in business of selling such product, and product is expected to and does reach user or consumer without substantial change in condition in which it is sold. (Davis v. Gibson Products Co., 68 N.J. 1, 342 A.2d 181, 184.)

Subpoena A subpoena is a command to appear at a certain time and place to give testimony upon a certain matter.

Subrogation The substitution of one person in the place of another with reference to a lawful claim, demand or right, so that he who is substituted succeeds to its rights, remedies, or securities. (Gerken v. Davidson Grocery Co., 57 Idaho 670, 69 P.2d 122, 126.) Insurance companies . . . generally have the right to step into the shoes of the party whom they compensate and sue any party whom the compensated party could have sued.

Tort A private or single wrong or injury, including action for bad faith breach of contract, for which the court will provide a remedy in the form of an action for damages. (K-mart Corp. v Ponsock, 103 Nev. 39, 732 P.2d 1364, 1368.)

Warranty A promise that a proposition of fact is true. An assurance by one party to an agreement of existence of fact upon which the other party may rely. It is intended precisely to relieve promisee of any duty to ascertain facts for himself, and amounts to promise to indemnify promisee for any loss if the fact warranted proves untrue. (Paccon, Inc. v. U.S., 399 F.2d 162, 166, 185 Ct.Cl. 24.)

Work-product rule Under this rule any notes, working papers, memoranda or similar materials prepared by an attorney in anticipation of litigation are protected from discovery. (Fed.R.Civ.Proc. 26(b)(3).)

Index